P9-AGV-036

THE HIDDEN HEART OF THE COSMOS

The Hidden Heart of the Cosmos is part of the Orbis Books Series "Ecology & Justice," edited by Mary Evelyn Tucker and John A. Grim, with Sean McDonagh and Leonardo Boff as advisors. Books in the Series seek to integrate an understanding of the Earth as an interconnected life system with concern to create just and sustainable relationships that benefit the entire Earth. *The Hidden Heart of the Cosmos* offers insights from contemporary cosmology that illuminate the search for the wisdom of the universe.

BRIAN SWIMME

THE HIDDEN HEART OF THE COSMOS

HUMANITY AND THE NEW STORY

ORBIS BOOKS
Maryknoll, New York 10545

The Catholic Foreign Mission Society of America (Maryknoll) recruits and trains people for overseas missionary service. Through Orbis Books, Maryknoll aims to foster the international dialogue that is essential to mission. The books published, however, reflect the opinions of their authors and are not meant to represent the official position of the society.

Published by Orbis Books, Maryknoll, New York, U.S.A.

Manufactured in the United States of America

Library of Congress Cataloging-in-Publication Data

Swimme, Brian.
The hidden heart of the cosmos : humanity and the new story / Brian Swimme.
p. cm.
Includes index.
ISBN 1-57075-058-0
1. Astronomy–Religious aspects–Christianity. 2. Cosmology. 3. Apologetics. I. Title.
BL253S85 1996
113–dc20 95-51969
CIP

Printed on recycled paper

To my parents,
Wayne and Jeanne Swimme

Contents

Preface		ix
Acknowledgments		xi
1.	The Way of Cosmology	1
2.	A Three-Hundred-Thousand-Year Lineage	8
3.	The Whirling Solar System	21
4.	Cosmology and Ecstasy	31
5.	The Sun as the Center	38
6.	Looking Down on the Milky Way	45
7.	The Large-Scale Structure of Space and Time	55
8.	The Story Came to Us	63
9.	Nighttime and Cosmic Rebirth	67
10.	The Place Where the Universe Began	75
11.	A Multiplicity of Centers	80
12.	Where Did the Universe Come From?	90
13.	All-Nourishing Abyss	97
14.	Einstein's Awakening	105
15.	The Center of the Cosmos	110
Index		113

Preface

From the beginning, humans have been pondering the ultimate nature of existence. Shamans and sages, philosophers and saints, rishis and rabbis and theologians, all in their various ways, have reflected on the deep and endlessly fascinating questions of existence. *The Hidden Heart of the Cosmos* wants to join that ancient tradition by asking the same questions, but from the perspective of contemporary science. Our reflections will begin with the universe as we now know it, a universe that began billions of years ago as hot dense plasma, that slowly unfolded into galaxies and stars, and that complexified further into living beings.

The question I consider in this present volume is perhaps the most ancient of all. "Where did it all come from? Where is the center of reality? Where is the heart or source of the universe? Where is that place where everything sprang forth into existence?" Relying on the discoveries of the modern scientific enterprise, and in particular of twentieth-century cosmology and quantum physics, we confront this perennial question not with any naive expectation that we will now answer with certitude questions which eluded our ancestors, but with the hope that we too might become just as engaged by the questions, and just as baffled and amazed by the answers.

In future volumes I hope to treat different questions, and will bring to these the central discoveries of the different sciences, but all of these meditations will begin within the context of the evolution of the cosmos, and will be guided by a single, unifying concern: "What does it mean to exist, as a human, in this vast unfolding universe? What is our role here? What is our destiny?"

Should such speculations originating from the world of science assist humanity in meeting and overcoming the challenges that beset us here at the eve of the third millennium, we will be justified in asserting that science as a whole now enters yet another phase of its journey. No longer simply the handmaiden of technology. No longer simply a materialistic study of reality. This new phase would be a science exploring the deep wisdom manifest in the universe. It may be too much to hope that science itself will become a new wisdom tradition, but perhaps we are already witnessing the creation of a comprehensive cosmology, one grounded in our contemporary understanding of the universe, and yet subtle enough to interact harmoniously with the more ancient cultural traditions.

Acknowledgments

I would like to thank Robert McDermott for creating the Philosophy, Cosmology and Consciousness Program at the California Institute of Integral Studies in San Francisco. It was there that these ideas took their present form. My thanks to Bruce Bochte and Marie Cantlon for help with the manuscript, and to my son Brian for his fine ink drawings. I particularly appreciated the editorial assistance of the Orbis team, including Mary Evelyn Tucker, John Grim, and Bill Burrows. Finally, my gratitude for the many people whose questions, criticisms, and suggestions have helped to shape every line.

1

The Way of Cosmology

The really surprising thing is that the news of the birthplace of the universe was always here. I mean that for as long as there have been humans on Earth there has always been this news of the universe's birthplace showering us day and night. This must be a central facet of human existence: to have the truth right in front of us and yet be unable to see it or recognize it.

In the case of the origin of the cosmos, the news is carried by particles of light – photons – and these particles have been here since the beginning. The problem is that the photons are too dim to be seen by the unassisted human eye. Just think of how many humans since the beginning of history wandered around bathed in the news of the origin of the universe, all of them simply unable to respond to what was quite literally right there before them.

The entire scientific enterprise can be characterized as the development of sensitivities and ideas necessary to become more fully aware of what is happening all around us. Seen in this perspective, the discovery of the birthplace of the universe is a four-million-year learning event. I say four million years because though it's difficult to identify where we should mark the beginning of humanity, one way to proceed is to take the origin as that moment when our ancestors first began walking on two legs. Some anthropologists prefer the moment when our ancestors first began using tools in a systematic way, around 2.5 million years ago. In any event, one has to stop and wonder over this drama where humans are wandering about, thinking, working, mating, suffering, even for millions of years, and throughout every moment of that long journey we have been suffused with the light from the beginning of time. The gift of the scientific venture is the capacity to see what was here all along. And the summary understanding achieved by this long process of inquiry can be stated so simply: *The birthplace of the universe, where existence first sprang forth, is fifteen billion light-years from the Earth.*

Most physicists regard this discovery of the birthplace of the universe as the most significant of the twentieth century. It was an enormously complex task involving thousands, and ultimately millions, of scientists. It was only by their careful and painstaking investigation of the universe – collecting data, developing new mathematical languages, debating the many interpretations – that the knowledge of the birthplace was finally secured. In this exploration of the cosmic center, transmission of information concerning the discoveries of modern science is of course a basic aim. But the greater challenge is to identify the meanings such discoveries have for human existence. These notes are rooted in a single, central conviction concerning our discovery of the birthplace of the universe, a conviction I will state baldly here and will return to throughout.

Four hundred years of modern science now reach a culminating moment with the discovery of the universe's birthplace. Certainly science's irruption in the sixteenth century was destructive not only of the European world-view, but of every traditional world-view with which science has had contact. But that destructive phase is now ending, and an integrative period begins. Even though science's violent rejection of every cultural and tribal tradition has been deleterious in the extreme, and even though one can appreciate the vehemence with which fundamentalist religions around the planet reject any compromise with modern secular scientific culture, the opportunity of our time is to integrate science's understanding of the universe with more ancient intuitions concerning the meaning and destiny of the human. The promise of this work is that through such an enterprise the human species as a whole will begin to embrace a common meaning and a coherent program of action.

One way to identify the significance of what is taking place is to say that science now enters its wisdom phase. This presentation on the new cosmology is part of our contemporary exploration of the wisdom within the great discoveries of the scientific enterprise as a whole. We are challenged here with understanding the significance of the human enterprise within an evolving universe. Upon our success in meeting this challenge rests the vitality of so much of the Earth Community, including the quality of life all future children will enjoy.

In approaching the discovery of the center of the cosmos, we need to bear in mind that it is different from the discovery of the Grand Canyon. If I were among the first paleo-Indians to see the Grand Canyon and wanted to share it with a friend, I would just take her there and point out the great red walls, the crumbling sun-baked rocks, the distant low hush of the hidden river.

Though the birthplace of the universe has physical dimensions and can be, in a sense, pointed at, it is a different kind of reality than the Grand Canyon, or the African Elephant, or the Sistine Chapel. After all, if the center were the kind of reality that could be simply pointed at and recognized in a glance, the discovery would not have required millions of scientists working ceaselessly over four centuries.

Some discoveries, precisely because they reveal an immense truth, are difficult to transmit. Imagine what it must have been like for that animal who first discovered conscious self-awareness. I think that discovery matches our own in both its significance and in its difficulty to transmit. We take self-awareness for granted now, but there was a moment in the past when conscious self-awareness appeared for the first time in the Earth Community. Some animal entered into the experience without understanding what was taking place, for no other animal had ever been in that mode of existence before. Although we have no way of knowing now where it began, let us imagine that it took place in an ancient form of the human, a pioneering hominid at the dawn of human history awakening to the awesome fact of awareness.

Suddenly this animal was not only conscious of the forest; suddenly she was also aware that she was aware. The animal could see the river and hear the mosquitoes, as always, but now she also noticed that it was she who was seeing and hearing. She, a self, an individual self, it was she who was aware of all this. The discovery of conscious self-awareness was the discovery by some self that her own self was enjoying all this.

If you were this primate, filled with a desire to tell your companions about your discovery, how would you go about doing so? Would pointing a finger work? Maybe. But what one should point at in order to indicate conscious self-awareness is hard to imagine. Perhaps you could bring your companions to a pond and point at their faces. That might work. But maybe not. Maybe something stranger, maybe a faked temper tantrum with angry back flips while baying at the Moon, maybe that would do the trick.

Though we don't know how these early animals communicated, we do know they came up with something that worked, for conscious self-awareness spread out, perhaps surfacing in several places before finally lodging as a pervasive feature in at least one group, *Homo sapiens*. As these early individuals entered conscious self-awareness, they did not awake to a thing, or to an event, but rather to a reality of each thing and of each event. They awoke to a reality that had in a sense always been at hand even though impossible to point out in any simplistic manner.

In this exploration I aim for a language that points at things such as stars and planets and oak trees, but that at another level carries an invitation for a journey to the center of the universe. Arrival at the center is simultaneously a physical and a nonphysical event. Arrival at the center involves understanding what we have learned about the universe, and then slowly appropriating this understanding into daily life.

Learning about the center of the universe is not always a pleasant experience. We will be discussing ideas that sometimes run counter to much of what we ordinarily take as true concerning the world. It's not always pleasant because it can be disturbing to deal with questions concerning the foundations of the universe. But there is no way around it. The discovery of the birthplace of the universe requires such an effort.

How convenient if truth were a shiny pebble we could pick off the beach and drop in our pockets. Stoop down, pluck it out, and it's ours. Certainly some truth can be grasped this way. "I had macaroni for lunch today" requires little effort for comprehension. But deep truths challenge us profoundly. To understand them demands a change in ourselves along with a creative leap of the imagination.

We need to remember that it took some of humanity's most brilliant representatives four centuries of uninterrupted investigation to discover the center of the universe. Even after the discovery, some of the most gifted scientists of all, in particular Albert Einstein, failed to grasp what it was they had discovered. If the discovery required so much and shocked some of our species' finest minds, we should prepare ourselves to put some effort into the work of grasping this new truth.

At one time, conscious self-awareness was alive in only a tiny fraction of the existing humans. But the destiny of the species was forever altered when this form of consciousness emerged in increasing numbers until in time it became the determining characteristic of humanity. So too with us. A new form of consciousness is beginning to emerge in a small slice of contemporary *Homo sapiens*. As with the early self-aware primates, we are astounded by the new awareness, and when we go to speak of it, we discover that we have no easy or established or efficient way of transmitting this mode of consciousness.

Perhaps in the future we will have invented a multiplicity of modes for communicating this awareness. But here I am presenting the approach I know best, the ancient process of cosmology. It consists simply of a sustained contemplation on the ways of the universe, with special attention, in our case, to the knowledge we have gained through the careful observation and meditation of the last four hundred years. By learning the ways of the universe and by reflecting upon them as they surface in the daily life of family and work and community, we take the first steps into a new form of human understanding and existence.

2

A Three-Hundred-Thousand-Year Lineage

Humans have been at this cosmological task a long time. I celebrate the new and startling results of our modern scientific investigation, but we need to remember that this enterprise of cosmology – the explorations of the origin, development, and destiny of the universe – has been with us from the beginning. Most spectacularly we have the cave paintings from as far back as twenty thousand years ago in southern Europe, when our ancestors crawled for days on their backs through the labyrinthine caves to gather in great underground vaults for their religious celebrations. Other artifacts of cosmological wonderment reach back perhaps forty thousand years, and some archeologists surmise that our human ancestors have been gathering in caves as far back as three hundred thousand years ago, haunted even then by the mysteries of the cosmos that filled them with such terror and delight.

None of the other animals of the world needs to engage in such reflections. They enter life and are given their basic relationships through the genetic programs that have been fashioned over millions of years. All except the human know their proper place in the world. But the human requires something different, something more than genetic codes. Humans require a cultural orientation. We are not given a fixed and final form to our orientation in life but must discover and deepen this orientation through the process of psychic development.

It is not because we have no answers to the question, "What does it mean to be human in this universe?" It is rather because we have so many different answers that we need to stop and wonder about the universe in order to sort out our right and fruitful relationships.

Conceivably for as long as three hundred thousand years, humans have huddled together in the night to ponder and to celebrate the mysteries of the universe in order to find their way through the Great World they inhabit. No matter what continent humans lived on, no matter what culture, no matter what era, the work of cosmology took place every year and every month and even every day – around the fire of the African plains, in the caves of the Eurasian forests, under the brilliant night sky of the Australian land mass, in the long houses of North America. There the people told the sacred stories of how the world came to be, of what the human brings into the universe, and of what it takes to live a noble life within the Great Holy that is the universe.

I say that every culture did this, but that of course is not exactly true. For we contemporary humans do not. Modern humanity seems to be the first culture to break with this primordial tradition of celebrating the mysteries of the universe. When we learn about other cultures who do, we feel a sense of superiority or nostalgia, depending on our evaluations of such cultures. If we regard the scientific enterprise as freeing us from earlier superstitions, we look with pity upon these primitives who devoted so much energy to teaching fantasies about the universe. On the other hand, if we think that earlier or other cultures knew something important about the universe that completely escapes our ways of knowing, we are left with a sense of sadness that such experiences are not possible for us today.

Modern industrial society does it differently. Questions of ultimate meaning and value are dealt with not in caves or on the open plains, but in the churches, mosques, and temples. Here each weekend billions of humans gather to reflect on their relationship with the divine – whether approached under traditions of God, Allah, Brahma, or the Great Spirit. In all these millions of weekly religious ceremonies, so essential to the health and spirituality of humanity as a whole, one will find such a diversity of religious celebrations, but only rarely will one find serious contemplation of these primal human questions *within the context of the actual universe, a universe of stars, topsoil, amphibians, and wetlands.*

Certainly in the world's scriptures and in the religious rituals performed each week there are powerful uses of such words as "water," "sky," "sun," and "rain." But these are employed in a symbolic, rather than a literal, sense. For instance, "water" might be used as a sign of the saving action of God; but it is not used to point to the hydrology of the Mississippi River. Evidence of this is found in our own minds: for us, the Mississippi River and God do not really intersect. God has to do with the gospel of love, with salvation, with care for the poor, with the drama of the Bible, whereas the phrase "the hydrology of the Mississippi River" connotes for us various "physical" things, like the H_2O molecules, or the dams of the Army Corps of Engineers, or the laws dealing with water rights – and all of it understood as separate from questions of God and ultimacy.

The result is that, within our religions, when we do ponder the deep questions of meaning in the universe, we do so in a context fixed in the time when the classical scriptures achieved their written form. We do not worship or contemplate in the context of the universe as we have come to know it over these last centuries, a context that includes the species diversity of the Appalachian mountains, the million-year development of the enveloping ecosystem, the intricate processes of the human genome, the stellar dynamics that gave birth to Earth five billion years ago, or anything else that is both specific and true concerning the Earth and universe. All of that – the Earth and universe as they are and as they actually function – is regarded as "science," something separate from questions of meaning and value that religions deal with.

Modern humans, instead of gathering in the caves or cathedrals to dance to poetry and music as a way of learning their place in the universe, sit in classrooms and study science. Certainly such education in the sciences is fundamental for the survival of humanity. The challenges that beset us today will grow ever more fierce for our children and their children, and we require the best science and technology we are capable of. But nowhere in science education – not in Europe or Asia or Africa or the Americas – is the fundamental role and meaning of the human and the universe treated in any significant manner. The ruling assumption is that science is concerned with facts, whereas meaning and purpose and value are the domain of religion.

The tragedy here is that our religions would remain true to their essence if they were to think and work within the larger context of the universe. It would not mean shrinking away from the central religious truths. On the contrary, expressed within the context of the dynamics of the developing universe, the essential truths of religion would find a far vaster and more profound form. The recasting would not be a compromise nor a diminution nor a belittlement; it would be a surprising and creative fulfillment, one whose significance goes beyond today's most optimistic evaluations of the value of religion.

During the modern era of humanity the ancient cosmological enterprise was broken apart, not by some accidental development but by the ways of thought rooted in the very core values of our modern world. The division between science and religion can be argued with libraries of philosophical texts from modern thinkers and defended with mountains of legal briefs. We are prevented from engaging in the cosmological enterprise precisely because the institutional processes of the modern world keep humans away from such questions.

But if humans, in order to become fully human, truly do need to ponder the universe to discover their place in nature, and if this three-hundred-thousand-year tradition is rooted in the requirements of our genetic make-up, then we will find our way to ideas concerning the proper human role in the universe one way or another. And if the institutions of education and religion have, for whatever well-defended reasons, decided to abdicate that role, someone somewhere else is going to step forward and provide it.

Where are we initiated into the universe? To answer we need to reflect on what our children experience over and over again, at night, in a setting similar to those children in the past who gathered in the caves and listened to the chant of the elders. If we think in terms of pure quantities of time the answer is immediate: the cave has been replaced with the television room and the chant with the advertisement. One could say that the chant has been replaced with the television *show,* but at the core of each show, driving the action, and determining whether or not the show will survive the season, is the advertisement. That is the essential message that will be there night after night and season after season. Television's *Bonanzas, Cheers,* and *Cosby* shows all come and go; the advertisement endures through every change.

What is the effect on our children? Before a child enters first grade science class, and before entering in any real way into our religious ceremonies, a child will have soaked in thirty thousand advertisements. The time our teenagers spend absorbing ads is more than their total stay in high school. None of us feels very good about this, but for the most part we just ignore it. It's background. It's just there, part of what's going on. We learned to accept it so long ago we hardly ever think about it anymore.

But imagine how different we would feel if we heard about a country that programmed its citizenry in its religious dogmas in such a manner. In fact, it was just such accounts concerning the leaders of the former Soviet Union that outraged us for decades, the thought that they would take young children and subject them to brainwashing in Soviet lies,

The fact that consumerism has become the dominant world-faith is largely invisible to us, so it is helpful to understand clearly that to hand our children over to the consumer culture is to place them in the care of the planet's most sophisticated religious preachers.

removing their natural feelings for their parents or for God or for the truth of history, and replacing these with the assumptions necessary for their dictatorship to continue its oppressive domination.

Immersed in the religion of consumerism, we are unable to take such comparisons seriously. We tell ourselves soothing clichés, such as the obvious fact that television ads are not put on by any political dictatorship. We tell ourselves that ads are simply the efforts of our corporations to get us interested in their various products. But as with any reality that we rarely pay any serious attention to, there may be a lot more going on there than we are aware of. Just the sheer amount of time we spend in the world of the ad suggests we might well devote a moment to examining that world more carefully.

The advertisers of course are not some bad persons with evil designs. They're just doing their job. On the other hand, we can also say that their primary concern is not explicitly the well-being of our children. Why should it be? Their objective is to create ads that are successful for their company, and this means to get the television viewer interested in their product. But already we can see that this is a less than desirable situation. After all, we parents demand that our children's teachers, to take just one example, should have our children's best interests foremost in mind. Such teachers will shape our children when they are young and vulnerable, so of course we want this shaping to be done only by people who care. So to hand over so much of our children's young lives to people who obviously do not have our children's well-being foremost in mind is at the very least questionable.

But at a deeper level, what we need to confront is the power of the advertiser to promulgate a world-view, a mini-cosmology, that is based upon dissatisfaction and craving.

One of the clichés for how to construct an ad captures the point succinctly: "An ad's job is to make them unhappy with what they have."

We rarely think of ads as being shaped by explicit world-views, and that precisely is why they are so effective. The last thing we want to think about as we're lying on the couch relaxing is the philosophy behind the ad. So as we soak it all up, it sinks down deep in our psyche. And if this takes place in the adult soul, imagine how much more damage is done in the psyches of our children, which have none of our protective cynicisms but which draw in the ad's imagery and message as if they were coming from a trusted parent or teacher.

Advertisers in the corporate world are of course offered lucrative recompense, and, with that financial draw, our corporations attract humans from the highest strata of IQs. And our best artistic talent. And any sports hero or movie star they want to buy. Combining so much brain power and social status with sophisticated electronic graphics and the most penetrating psychological techniques, these teams of highly intelligent adults descend upon all of us, even upon children not yet in school, with the simple desire to create in us a dissatisfaction for our lives and a craving for yet another consumer product. It's hard to imagine any child having the capacities necessary to survive such a lopsided contest, especially when it's carried out ten thousand times a year, with no cultural condom capable of blocking out the consumerism virus. Could even one child in the whole world endure that onslaught and come out intact? Extremely doubtful. Put it all together and you can see why it's no great mystery that consumerism has become the dominant world faith of every continent of the planet today.

The point I wish to make is not just that our children are such easy prey. It's not just that the rushing river of advertisements determines the sorts of shoes our children desire, the sorts of clothes and toys and games and sugar cereals that they must have. It's not just the unhappiness they are left with whenever they cannot have such commodities, an unhappiness that in many cases leads to aggressive violence of the worst kinds in order to obtain by force what their parents will not or cannot give them. All of this is of great concern, but the point I wish to focus on here has to do with the question of how we are initiated into a world.

Advertisements are where our children receive their cosmology, their basic grasp of the world's meaning, which amounts to their primary religious faith, though unrecognized as such. I use the word "faith" here to mean cosmology on the personal level. Faith is that which a person holds to be the hard-boiled truth about reality. The advertisement is our culture's primary vehicle for providing our children with their personal cosmologies. As this awful fact sinks into awareness, the first healthy response is one of denial. It is just too horrible to think that we live in a culture that has replaced authentic spiritual development with the advertisement's crass materialism. And yet when one compares the pitiful efforts we employ for moral development with the colossal and frenzied energies we pour into advertising, it is like comparing a high school football game with World War II. Nothing that happens in one hour on the weekend makes the slightest dent in the strategic bombing taking place day and night fifty-two weeks of the year.

Perhaps the more recalcitrant children will require upward of a hundred thousand ads before they cave in and accept consumerism's basic world-view. But eventually we all get the message. It's a simple cosmology, told with great effect and delivered a billion times each day not only to Americans of course but to nearly everyone in the planetary reach of the ad: *humans exist to work at jobs, to earn money, to get stuff.* The image of the ideal human is also deeply set in our minds by the unending preachments of the ad. The ideal is not Jesus or Socrates. Forget all about Rachel Carson or Confucius or Martin Luther King, Jr., and all their suffering and love and wisdom. In the propaganda of the ad the ideal people, the fully human humans, are relaxed and carefree – drinking Pepsis around a pool – unencumbered by powerful ideas concerning the nature of goodness, undisturbed by visions of suffering that could be alleviated if humans were committed to justice. None of that ever appears. In the religion of the ad the task of civilization is much simpler. The ultimate meaning for human existence is getting all this stuff. That's paradise. And the meaning of the Earth? Premanufactured consumer stuff.

I have mentioned only television here, but of course that is simply one part of the program. To wade into a fuller awareness we need bring to mind our roadside billboards, the backs of cereal boxes, the fifty thousand magazines crammed with glossy pitches, the lunch boxes wrapped with toy advertisements, the trillion radio commercials, the come-ons piped into videoprograms, the seductions pouring into the telephone receiver when we're put on hold, the corporate logos stitched into our clothes and paraded everywhere and so on and so on. Literally everywhere on Earth, the advertising continues its goal of becoming omnipresent, even entering into space on the surfaces of our capsules.

None of what I have said here concerning ads and their effects on children will be news to those educators who for decades have been lamenting this oppressive situation in America. But I bring up the issue for two reasons.

The fact that consumerism has become the dominant world faith is largely invisible to us, so it is helpful to understand clearly that to hand our children over to the consumer culture is to place them in the care of the planet's most sophisticated religious preachers. If those bizarre cults we read about in the papers used even one-tenth of 1 percent of the dazzling deceit of our advertisers, they would be hounded by the federal justice department and thrown into jail straightaway. But in American and European and Japanese society, and increasingly everywhere else, we are so blinded by the all-encompassing propaganda we never think to confront the advertisers and demand they cease. On the contrary, as if cult members ourselves, we pay them lucrative salaries and hand over our children in the bargain.

The second reason for bringing up the advertisement's hold on us has to do with my fundamental aim in presenting the new cosmology. If we come to an awareness of the way in which the materialism of the advertisement is our culture's primary way for shaping our children, and if we find this unacceptable, we are left with the task of inventing new ways of introducing our children and our teenagers and our young adults and our middle-aged adults and our older adults to the universe. These notes on the new cosmology are grounded in our contemporary understanding of the universe and nourished by our more ancient spiritual convictions concerning its meaning. These notes then are a first step out of the religion of consumerism and into a way of life based upon the conviction that we live within a sacred universe.

3

The Whirling Solar System

The earliest cosmologies regarded Earth as the center of the cosmos. So common is this idea among the peoples of the various continents, that the notion of Earth's centrality in the universe might be considered a primal cosmological intuition of humanity. In one particular expression of this, the western European, we can still enjoy a supreme expression of the geocentric cosmology splashed across the ceiling of the Sistine Chapel. There Michelangelo tells the story of the creation in nine panels. The story begins with God fashioning the Earth and ends with the human drama of sin and salvation taking place at the center of the universe. Other classical civilizations were built upon similar geocentric conceptions, all of which differ radically from our current view of the universe.

The modern scientific endeavor began with the destruction of humanity's common geocentric cosmology. In 1543, Nicolaus Copernicus, an obscure Polish astronomer, announced that the Sun was the center of the world. "Announced" is not the best word. People were surely announcing all sorts of things back then, just like today. Some people were announcing that the Earth was flat. Others announced that the Sun was Ra. The difference in Copernicus's case was that he provided a book with his announcement, and the book provided a way to understand that the Sun was the center around which the planets moved.

What we need to appreciate here is the daring of the whole operation. For maybe hundreds of thousands of years humans took as obvious that the Earth was the center of the universe. Those earlier humans would have been deeply confused by any suggestion that the Sun resided at the center and that the Earth was spinning around it. Such an idea, to say the least, is far from obvious.

For the naive or uncritical mind, the Sun is this hot thing up in the sky that travels around the Earth every day. We can't tell how big it is, but it couldn't be that big, because you can block it out entirely with just your thumb. The Earth, in contrast, is the whole world! It's a place of great oceans and tidal waves and vast mountain ranges and frightening hurricanes! Of blizzards and avalanches!

For the early, naive human consciousness, the Earth was obviously the most stable place in this universe. It stood still year after year while all the seasons came and went and the stars and planets and sun and moon whirled about it. To move from this naive and entirely natural understanding to a view that is profoundly "unnatural" and counterintuitive was certainly a strange and dangerous step to take.

The greatness of Copernicus is that with his book he provided a process by which the most advanced thinkers of Europe could come to grasp this new, subtle, disturbing, and amazing truth – the Sun resides in the center of the solar system while the Earth and Mars and Jupiter and all the planets circle about it. To gauge the greatness of Copernicus and the other founders of the modern scientific understanding one needs to appreciate their achievements from the perspective of biological evolution. From the time it began fashioning its first tools, humanity required two million years to construct a way to see through the naive and natural assumptions about the Earth and Sun in order to arrive at an understanding more firmly rooted in the actual operations of the universe. These early modern scientists then didn't just offer yet another cultural idea; they broke with a two-million-year tradition in human knowing, for they introduced knowledge that was so far from common sense it even flatly opposed what seemed unquestionably true.

The monumental nature of Copernicus's breakthrough can be appreciated directly. Though all of us now accept the fact that the Earth moves around the Sun, this truth has been actually appropriated *in a bodily way* by only a tiny segment of humanity. The cultural storms about the Sun's centrality have come and gone, and yet most of us watching a sunset have an experience similar to what medieval people experienced when they watched a sunset, which was the same as what people in the classical civilizations experienced, as well as those back still further, in the Neolithic and Paleolithic eras. In fact, this common experience of sunset is just what any primate would have experienced anytime since the very beginning of primate life, seventy million years ago. We all watch the Sun push down and then drop below the dark, unmoving horizon. If asked afterward what we were doing, we say, "I was watching the Sun go down." But if asked to explain what happened, we say, "Well, you know, the Earth is spinning so it just appears as if the Sun is going down."

In order to move from ordinary consciousness to a new

kind of planetary wisdom, we need a transformation of our experience that takes place concurrently with our acquisition of knowledge. It is not enough simply to learn more facts and knowledge about the universe. Something much deeper and more difficult is necessary. This new challenge is difficult precisely because of the counterintuitive nature of scientific discoveries. Science arrives at truths that are not part of our genetic inheritance, and thus they often appear strange and unnatural. But so long as such truths are left to dangle outside as abstractions, we are condemned to live a split life.

If we consider the situation under question we can appreciate the whole difficulty in concise form. Our primate perceptual habits of consciousness cause us to see the Sun "going down." Our observational and theoretical studies, on the other hand, have discovered the Earth's revolution about the Sun. What is needed here is a transformative process where one can learn to see and to feel the world in a way congruent with what is actually happening. Such a transformation would enable one to transcend the split modern condition of experiencing the world one way, while knowing the truth of the world is otherwise.

I do not know of any science department in the American system of higher education where a change of perception is a primary aim of the curriculum. Our focus of course has been dominated by the central task of accumulating and producing knowledge. Learning to actually experience a dynamic evolving universe does occur, but always in a haphazard manner as a by-product of the primary focus. What I am suggesting is that such a transformation of one's subjectivity might become an explicit goal in the next millennium, not to be considered as a replacement but as a completion of the traditional goal of knowledge acquisition.

My aim here is not simply to hand over information as if I were passing on a sheaf of papers from me to you. My aim is to present the birthplace of the universe in a way that invites you to participate in an inner transformation. It would be a great thing if a person learned the facts of the new story. But even greater would be to take the first steps into *living* the new story. We study the story primarily in order to live the story.

A creative event in our ancestry provides an image for the challenge we are confronting here. If you go back far enough in time you find that all mammals had eyes on each side of their heads. Whatever they chose to look at they saw with one eye at a time, in much the same way as a horse or a rhino today. But then one day a particular group of these mammals found themselves living in the trees of a forest and things began to change.

In the new challenge of a thick tangle of branches and vines, their eyes migrated together in a long and fitful process that eventuated finally into stereoscopic vision and its depth perception. These new mammals were the primates, and their ability to judge with great accuracy the distance from hand to branch assisted them in their triumphant radiation throughout the forests of the continents. If we consider the primate line as a whole, we can think of their journey as a single process whereby they learned to adjust to the world that surrounded them in a way that was more successful than what they had experienced before.

We are similarly challenged today. We are six billion humans, and we need to learn to live with one another and with all the other ten million species of life in a mutually enhancing way. We fail at the present time precisely because we fail to see and understand what it is that surrounds us. In order to awake we need nothing like the anatomical changes that enabled the primates to survive but, rather, a new mind and a new story that will enable us to inhabit successfully this quantum evolutionary cosmos. We need to learn to experience directly the more subtle complexities of the seamless whole that is nature or cosmos, which includes the events of our moments as well as the great events of the past and the unformed events following from our moment now. When we learn to experience our world in such a manner, we will have crossed into a new way of being human just as the primates crossed into a new way of being mammals.

In order to take a beginning step in this regard, I'd like to focus on this one particular experience of a sunset. Any person who wishes to can transform her perceptual habits and can learn to see, in a direct experiential way, the Earth rotating away from the Sun. The simplest way to accomplish this is to go out half an hour before "sunset" at a time when Venus is low on the horizon. It is helpful if another planet is also visible, such as Jupiter or Mars, but that is not necessary. A last ingredient for this process is a child. If you can manage to bring along a child, she will probably get there first, and her glow of discovery will assist you in making your own advance.

Begin by focusing your attention on Venus, and as you do so, keep the model of the solar system in mind as a way of organizing your experience. So, either by speaking out loud or by considering the situation in your thoughts, review the basic facts: "Venus is 65 million miles from the Sun, about a third closer than the Earth, which is 93 million miles from the Sun. And there, higher up in my field of vision, is Jupiter, 480 million miles from the Sun. All three of us are moving in a single plane around the Sun." And so forth. The actual numbers are not even necessary here. What needs to be kept in mind is the simple fact that we are dealing with great distances, and that – concerning the three planets – Venus is closest to the Sun, then Earth, then Jupiter, and all three move in a plane about the Sun.

Simply by focusing on the experience and viewing it through the theoretical model of the solar system's form, there comes a wonderful moment when you enter into it all at once: you feel in an experiential, imaginative, and direct way the Earth slowly turning away from the Sun. You have a sense of the plane in which the planets move, and even a beginning recognition of the great distance to Venus. You will also feel, and perhaps for the first time in your life, the immensity of the Earth as it rolls away from the great Sun. It happens in a flash. A single surprising shudder passes through you and you realize you are standing on the back of something like a cosmic whale, one that is slowly rotating its great bulk on the surface of an unseen ocean.

It is true that, soon afterward, we snap right back into our everyday way of experiencing the world. But if even for a moment you enter this larger experience of the world, you will be able to enter it again more easily in the future. The primary gateways are dawn and dusk, but as you grow in competence you can learn to experience yourself on the whirling Earth amid the enveloping solar system at any time and at any place. In each such moment you remain of course an individual person on the planet but you become as well a living planet encircling a star.

As you learn to feel directly the immensity of the Earth rotating away from the Sun, you can then take the further step of feeling the Earth swinging around the Sun.

Though we say, "The Earth revolves around the Sun," the truth is slightly different. The Sun is also moving, though not as much as the Earth is moving. An image that can help fix this in mind is that of the hammer throw in Olympic competition. Here a person swings a hammer around and around and then flings it through the air. A careful examination of the movement would show that the human is not at the fixed center of the spinning motion but that the human and the disk both revolve about their common center of gravity. The human doesn't move in as big a circle as the hammer because the human weighs much more than the hammer; and in a similar way, the Sun does not move anywhere near as much as the Earth, because the Sun is so much heavier. Nevertheless, the Sun does move in a tiny circle as both the Sun and Earth revolve about their common center of gravity. The point of their center of gravity is inside the Sun, but is not at the Sun's center.

If, again at dawn or dusk, you contemplate the same setting, with Venus in the sky and a child at your side, you can get your first real taste of the Earth's movement around the Sun and the Sun's tiny wobble in response.

One additional fact will aid your contemplation here. The size of the Sun is approximately a million times the size of the Earth. Thus, if you start your contemplation with a sense of Earth's immensity, and you now imagine that hot bright being on the horizon as containing a million earths, you can begin to feel the way in which the massiveness of the Sun whips the Earth and all the other planets through their annual arcs. The crucial step here is to awaken to the fact of the Sun's gravitational power. The Earth is one immense planet, and it is being whipped around the Sun *by the power of the Sun.* This is something the Sun is *doing* in every instant of every day. We are held by the Sun. If the Sun released us from our bond with it, we would sail off into deep space.

As before, a new awareness will come in a sudden shift where a door opens and you feel yourself sliding into an unsuspected and disorienting awareness. It is disorienting not in the sense of an irritated confusion – for the experience is not at all irritating but on the contrary is usually breathtaking; it is disorienting in the sense of a bottom dropping away, as if for the first time in your life you have closed your eyes and leapt into a body of cool water and are suddenly turning about weightless without toes or fingers touching any ground.

Words are pathetically inadequate to convey this cosmological experience. Modern English as a living language was created over the last five centuries by humans who did not have this particular experience, so how likely is it that English would contain the verbs, adjectives, metaphors, and rhetorical images necessary to convey it? There is no linguistic formulation that would make unnecessary your direct contact here. To fully understand, one has to sit down and wait for the universe to enter.

But if you do so, you will become one of a very small number of humans *who actually live in the solar system.* Most humans live not on the Earth that rotates and revolves about the Sun. We live rather in a fantasy that regards the Earth as a fixed place, where the ground is always stationary; that regards the planet as somehow resting on a great slab of cement. But to contemplate the solar system until you feel the great Earth turning away from the Sun and until you feel this immense planet being swung around its massive cosmic partner is to touch an ocean of wonder as you take a first step into inhabiting the actual universe and solar system and Earth.

Of course, all of this is very flimsy. This new cosmological orientation, even if deeply felt, will soon crumble as we are buffeted back into the nihilistic materialism dominant in our society. That's just where we find ourselves today, and we have to start somewhere. But it is helpful to picture how, in the next millennium, each day might be filled with opportunities to experience the truth of our endlessly complexifying universe. As you walk through your day tomorrow, just imagine each ad you see replaced by a work of art or a ritual or an educational process designed to establish you in the creativity of an unfolding universe. It will be only in such a culture that this new orientation becomes ordinary and natural.

For our lives today, our tasting of the universe will be sporadic and brief. But even if such encounters as I've described are ephemeral, they will nevertheless enable you to experience your self and your world, perhaps for the first time in your life, as soaring around a star that floats in the vast ocean of the cosmos.

4

COSMOLOGY AND ECSTASY

Since my educational training and professional work are in science, I inevitably share all the shortcomings of my tribe, and thus in particular I overemphasize the intellectual side of this transformative experience. If the task of initiating humans into the universe were solely intellectual I could – as in any science class – simply teach the facts and theories of the universe. But to treat the radical discovery of an unfolding universe as if it were another fact similar to, say, the melting temperature of lead, is to miss the whole purpose of cosmology.

Cosmology, though it is consonant with science, is not science. Cosmology is a wisdom tradition drawing upon not just science but religion and art and philosophy. Its principal aim is not the gathering of facts and theories but the transformation of the human. In the situation discussed in the previous chapter, we would say that science aims at an understanding of the Earth's rotational and revolutionary movements around the Sun, while cosmology aims at *embedding a human being in the numinous dynamics of our solar system*.

This experience of feeling yourself embedded in the whirling solar system is certainly not solely cerebral. There is always a strong emotional and bodily experience in any entrance into the universe. Such moments are often even tinged with a kind of ecstasy. And unless this full-bodied experience is pursued, we are settling for abstract understanding rather than a full initiation into the universe.

I mention this matter of the aesthetic and affective dimensions for a couple reasons, the first of which is to encourage further explorations in this endeavor. Since the modern system of education does not focus primarily on the transformation of the subjectivity of the student, we as a culture do not have any carefully worked-out educational practices we can employ. All classical civilizations developed highly intricate and effective methods for initiating their young into their culture's world-view; and the modern industrial West does as well, using advertisements to inculcate consumerism. But in the case of the new cosmology of a universe that came into existence fifteen billion years ago and has been complexifying ever since, there has not been time enough to create a corresponding civilizational discipline. Until the creative artists and educators and thinkers attend to this challenge we will continue to lose ourselves and the future generations to consumerism. Daring and bold initiatives are necessary in this endeavor. There are no experts here, and we must be happy in that, since the time calls not for expertise but genius.

The second reason for bringing up the aesthetic and affective component is to prepare anyone interested in this work of cosmological education for the sort of response she will receive. I can perhaps indicate what I mean by referring to my own experiences.

When I first began discussing some of the feelings that inevitably accompany any entrance into this new cosmological orientation, I was often asked if I used drugs. Such questions irritated and even threatened me. My initial unthinking response was that the questioner was suggesting everything I had to say could be reduced to a simple drug experience. Making the whole topic even harder to deal with was the destruction that drugs were causing for individuals and families and communities everywhere in our country. It was just too much for me to think about rationally.

But slowly, over time, as such questions continued, I came to realize that there is more going on here than I was aware of at first. After pondering the matter from a number of different angles I eventually came to adopt an entirely different hypothesis concerning the cosmology-drug relationship.

I now think alcohol and drugs are an intrinsic feature of the consumerism life-style. To state my conviction as bluntly as possible, I think that hoping for a consumer society without drug abuse is as pointless as hoping for a car without axle grease. Drugs of one form or another are necessary for the consumerism to continue.

It comes down to the psychic condition of a person in a consumer world. Consumerism is based on the assumption that the universe is a collection of dead objects. It is for this reason that depression is a regular feature in every consumer society. When humans find themselves surrounded by nothing but objects, the response is always one of loneliness, and here at the end of the second millennium we are swamped by a vast loneliness that has soaked into every stratum of our society. It is a sad though arguable fact that in the history of civilization there has never been so much loneliness in any society anywhere compared to that in contemporary industrial consumer society.

But isolation and alienation are profoundly false states of mind. We were born out of the Earth Community and its infinite creativity and delight and adventure. Our natural state is intimacy within the encompassing community. Our natural genetic inheritance presents us with the possibility of forming deeply bonded relationships throughout all ten million species of life as well as throughout the nonliving components of the universe. Any ultimate separation from this larger and enveloping community is impossible, and any ideology that proposes that the universe is nothing but a collection of pre-consumer items is going to be maintained only at a terrible price.

Sometimes, seen from the outside, the whole thing almost seems to work. Four years at the university and an engineering job to make heat-transfer modules for Chevy station wagons. And there's the daily commute on asphalt and breakfast in the kitchen on linoleum and the kids at school in Formica. Life becomes the blare of the radio and the DMV forms to fill out and the health insurance claims that need to be filed and somewhere in there a moment to glance at another family member usually being radiated by the TV. Finally on the weekends we're free from the roads and cars and traffic and offices and assembly lines and maybe that will make it all worthwhile, but two days dominated by electronic entertainment, which can of course be stretched out into Monday Night Football, is no real escape from having to face the bleak week all over again.

Consumerism is a prison whose walls and bars are the items advertised everywhere. We dedicate ourselves to getting the objects so that we can live encased by them. For most humans, even in the best of consumer circumstances, such a way of life proves unsatisfying to the core. It is simply not human finally to live a life sealed off from all conscious contact with those powers at work throughout the Earth and universe and within every one of our cells. So intolerable is this sense of being out of it, of being left out, of being without any central meaning for the world, we will resort to any route to ease the pain. And the quick and mindless way of transforming this deprived state of being is to ingest mind-altering chemicals that dissolve the thin veneer of consumer culture and bring one swimming into the primary processes at last. Thus, if only for a moment, and sometimes at a horrendous cost to self, family, friends, and community, one can be at home again in the great flood of beauty.

It really does not matter how big we make the consumer wall that separates us from the universe; it is always going to be made of sand. Even if we use shovels and hire a crew and heap it up into a great mound, the ocean tides are going to roll on in and curl up into immense breakers that obliterate them once again. No matter how many after-school drug programs we volunteer for, and no matter how many new jails we construct, and no matter how many new police we employ, and no matter how many drug dealers we hunt down and incarcerate, the number of lives ruined by drug abuse will continue to grow so long as we try to convince ourselves that we can live in an asphalt-girded, machine-dominated, biologically destructive, and spiritually desiccated consumer society out of touch with the numinous powers pervading each being in the universe.

If three hundred thousand years of human development

is any guide, it's a simple choice. We need to put our energy into inventing new cultural forms for initiating ourselves into an ecstatic sense of involvement with the community of beings that is the very universe. If we refuse to devote ourselves to this work, we'll just have to make the necessary adjustments to deal with the river of misery flowing out of that perverse way of life yoking insatiable greed with drugs known as consumerism.

Cosmology, when it is alive and healthy in a culture, evokes in the human a deep zest for life, a zest that is satisfying and revivifying, for it provides the psychic energy necessary to begin each day with joy. In the process of cosmological initiation as practiced by humans for three hundred thousand years, the pain of loneliness and isolation is replaced by the joy of bonded relationship.

But in contemporary America, when an audience hears about such feelings of joy and even ecstasy, the only way it can make sense of them is to think in terms of drugs. Somewhere in our past we shucked off that old business of initiating ourselves and our young into a sacred universe. And precisely because we do not have a cosmological tradition in contemporary America we are forced to conclude that any deep feelings evoked by cosmology must be somehow drug-induced.

In the ancient world, the ones who introduced the young to the universe were the shamans and the magicians. These individuals were highly skilled in the ways of shaping human consciousness, and those individuals today who are drawn to the cosmological enterprise need to take note. Those who would involve themselves with the new cosmology need a similar wisdom, for all the psychic energy now imprisoned by the fragile bars of consumerism will be increasingly drawn away and poured out into the ecstatic life within a new cosmological orientation. One should not go blindly into this enterprise. One needs to know that to work with these new educational processes is to work with an immense reservoir of human energy.

5

THE SUN AS THE CENTER

Some scholars, and even otherwise respectable scientists, make the mistake of claiming that twentieth-century science has proven Copernicus wrong. Not that the medieval geocentric model of the universe was correct; rather, both that model and Copernicus's heliocentric model are false. The argument goes as follows: Copernicus thought that the Sun was the center of the world, but we now know that the Sun is only one of trillions of stars in the universe, none of which is qualitatively different from all the rest. So Copernicus's notion couldn't have been correct.

In a superficial sense, this argument is flawless, for certainly it is true that Copernicus had no idea how vast the universe is. His data restricted his thinking to a much, much smaller cosmos, and he would have been astounded to learn the dimensions of the universe we now deal with. But the more important point, one that is missed entirely by this line of reasoning, is that Copernicus discovered *that the Sun is the center of the dynamics of our solar system*. Nothing learned about the universe since the time of his death changes this truth in the least. Though with Newton and Einstein and Feynman and McClintock we have enlarged and deepened our scientific understanding in tremendous ways, Copernicus's radical insight concerning the centrality of the Sun lives undisturbed inside all our contemporary theories about the universe.

On the contrary, as I would like to explore in this chapter, in the five centuries since Copernicus's death, our ongoing meditation on the universe has led us to an even deeper understanding of how the Sun is at the center of the solar system. We have already touched upon this with the issue of the size of the Sun. The Earth is just the tiniest fraction of the Sun's matter – only a millionth of the Sun's volume. The Earth and the other planets then are just wisps silently sailing through a space suffused with our star's brightness. A human with a cosmological education should learn that truth and should feel it directly in a bodily and imaginative way.

But the discovery I would like to examine here has to do with energy. Using the detailed understanding of atomic and nuclear physics discovered and developed in our own century, we have learned something never even suspected by the greatest thinkers of all human history, including Copernicus himself, and Galileo, and Aristotle, and Confucius, and Plato, and all the others. The Sun, each second, transforms four million tons of itself into light. Each second a huge chunk of the Sun vanishes into radiant energy that soars away in all directions. In our own experience we have perhaps watched candles burn down or have seen wood consumed by flames leaving behind only ashes, but nothing in all our human experience compares to this preternatural blaze that engulfs oceans of matter each day.

The ancient Greeks conveyed their deepest truths through poetry and myth, and thus bequeathed to us the stories of Apollo and Hephaestus and Aphrodite and Athena and

Zeus. Were we at that same stage of human consciousness, such myths would enable us, as they enabled the Greeks, to enter into a rich relationship with the powers in the universe. But they no longer work in the same way for us. Precisely because such stories are false in any literal sense concerning their descriptions of the universe, psychologists such as C. G. Jung and others have rescued them from the scrap heap of history by explaining to us their psychological relevance. We are thus condemned by our scientific knowledge to regard them, ultimately, as "myths," as fictions, as clever ways to teach psychological truths.

Thus, when we come to the fact of the Sun's massive transformation into energy, we are stymied. We have no myths or poetry that enable us to take this in. It all collapses down to just another fact from the new sciences that sits there and stares at us. And it is so alien, this profligate and monstrous discharge of energy. If anything, we distance ourselves from it. It is yet another forbidding truth about the inhuman universe, and we unconsciously commit ourselves to the modern disaster of sealing ourselves away from the universe.

Here is yet another gateway through which the cosmological imagination walks toward a new synthesis of science and religion. In the case of the Sun, we have a new understanding of the cosmological meaning of sacrifice. The Sun is, with each second, giving itself over to become energy that we, with every meal, partake of. We so rarely reflect on this basic truth from biology, and yet its spiritual significance is supreme. The Sun converts itself into a flow of energy that photosynthesis changes into plants that are consumed by animals. So for four million years, humans have been feasting on the Sun's energy stored in the form of wheat or maize

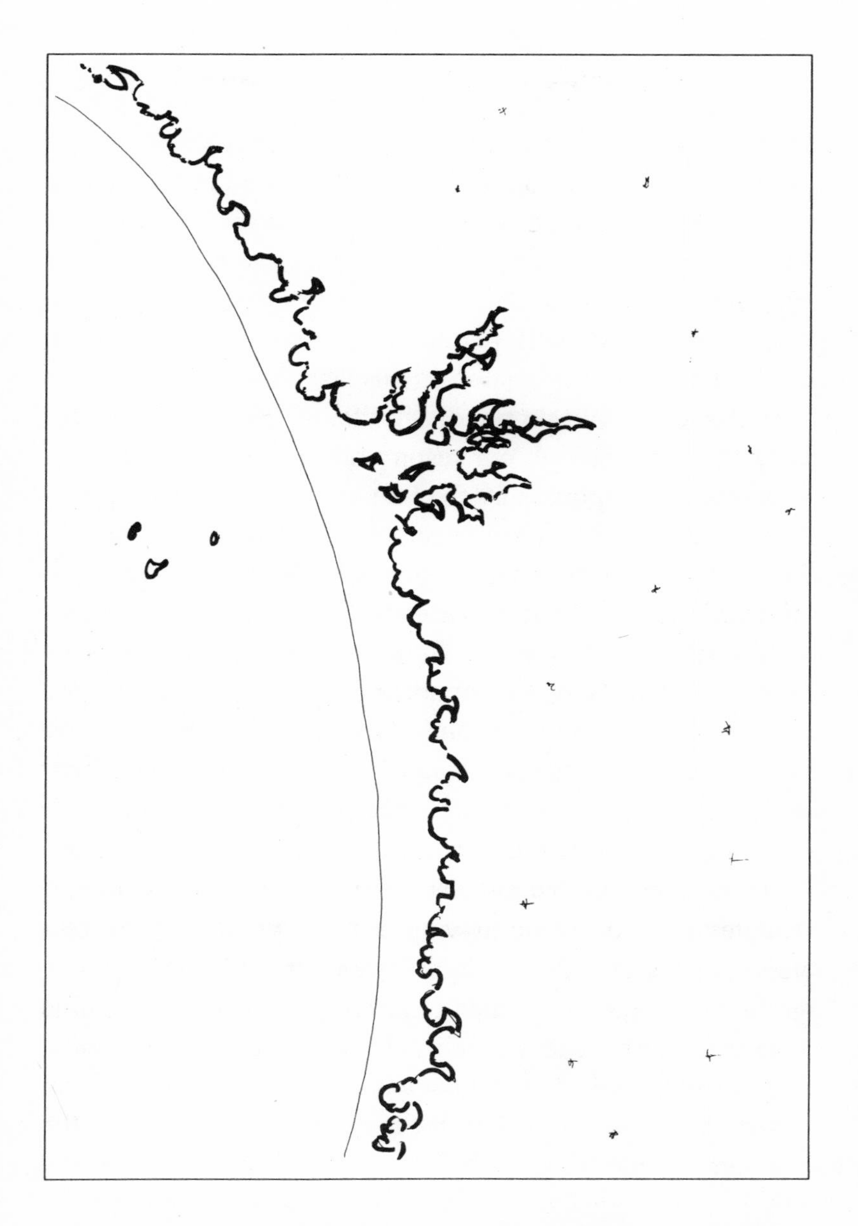

The Sun, in each second, transforms four million tons of itself into light.

or reindeer as each day the Sun dies as Sun and is reborn as the vitality of Earth. And those solar flares are in fact the very power of the vast human enterprise. And every child of ours needs to learn the simple truth: she is the energy of the Sun. And we adults should organize things so her face shines with the same radiant joy.

During the modern period when materialism came to dominate, such a suggestion as this last would be rejected as "mere poetry." We simply did not know that the actual energy coursing through our respiratory and nervous systems was bestowed upon us by the Sun and that our own vitality is a natural evolutionary development of the Sun's vitality. So instead of introducing our young to the Sun, we cut them off from the Sun. That is, instead of awakening this primordial relationship that would shine on the child's face with the radiance of the Sun, we unknowingly and tragically snuffed it out. They were left with our own convictions, that the universe was a collection of dead objects, and so it went from generation to generation throughout the modern world.

In the cosmology of the new millennium the Sun's extravagant bestowal of energy can be regarded as a spectacular manifestation of an underlying impulse pervading the universe. In the star this impulse reveals itself in the ongoing giveaway of energy. In the human heart it is felt as the urge to devote one's life to the well-being of the larger community.

In a culture where cosmology is living, children are taught by the Sun and Moon, by the rainfall and starlight, by the salmon run and the periwinkle's hideout. It has been so long since we moderns have lived in such a world, it is difficult to picture, but we can just now begin to imagine what it might be like for our children, or for our children's children.

They will wake up a few moments before dawn and go out into the gray light. As they're yawning away the last of their sleep and as the Earth slowly rotates back into the great cone of light from the Sun, they will hear the story of the Sun's gift. How five billion years ago the hydrogen atoms, created at the birth of the universe, came together to form our great Sun that now pours out this same primordial energy and has done so from the beginning of time. How some of this sunlight is gathered up by the Earth to swim in the oceans and to sing in the forests. And how some of this has been drawn into the human venture, so that human beings themselves are able to stand there, are able to yawn, are able to think only because coursing through their blood lines are molecules energized by the Sun.

And then they will hear the simple truth about the necessity of such a bestowal. If we burn brightly today it is only because this same energy was burning brightly as the Sun a month ago. Even as we take a single breath our energy dissipates and we need to be replenished all over again by the Sun's gift of fire. If the Sun were suddenly to stop transforming itself into energy, all the plants would die as the Earth's temperature plummeted hundreds of degrees below zero. In our veins and flesh, all the heat-giving molecules would go cold without replenishment as we and everything else became hard as frozen dust.

The Sun's story will find its climax in a story from the human family of those men and women whose lives manifested the same generosity and whose sacrifice enabled others to reach fulfillment. If through the ages the various cultures have admired such people who poured out their creative energies so that others might live, we were only intuitively recognizing that such humans were true to the nature of the energy that filled them.

Human generosity is possible only because at the center of the solar system a magnificent stellar generosity pours forth free energy day and night without stop and without complaint and without the slightest hesitation. This is the way of the universe. This is the way of life. And this is the way in which each of us joins this cosmological lineage when we accept the Sun's gift of energy and transform it into creative action that will enable the community to flourish.

Of course, over the years, as the Sun's story is repeated even in its various forms, there will be a good deal of repetition, and the listeners will sometimes be bored and distracted. This is to be expected. This is not entertainment but education. And moral education in particular rests upon holding in mind, over long periods of time, the magnificent achievements of fifteen billion years of creativity.

By reminding ourselves of the possibilities of true greatness and true nobility of spirit, we excite the energies necessary to achieve our true fulfillment. Then the challenge of moral and spiritual achievement is not something dealt with an hour on the weekend. Then the task of transformation is the way we start each day as we remind ourselves of the revelation that is the Sun.

Through repetition and through years of deepening, our children or our children's children will be provided a way to escape the lures of so much deceit, and greed, and hatred, and self-doubt, for they will begin each morning and live each day inside the simple truth: a gorgeous living Earth drifts light as a feather around the great roaring generosity of the Sun.

6

Looking Down on the Milky Way

Even if the discovery of the birthplace of the universe is the greatest of the twentieth century or of all time, it is meaningless until it comes alive within us. The discovery itself was not the result of an accidental or automatic action, but came through the sustained effort of millions of humans. So too with its meaning. It is easy for anyone to became momentarily fascinated or titillated with the wild data of the new story of the universe, but it is another thing altogether to absorb this over time into the center of one's being. Facts by themselves are not enough; what is needed is embodiment. What is needed is a transformation from the form of the humanity of today into forms of humanity congruent with the ways of the universe. Such a re-education will take place only in those individuals who have the courage, imagination, and energy necessary for the journey.

To study the new cosmology is to have your consciousness transformed. One of the more depressing insights in this transformation is to realize how efficient we have been in sealing ourselves and our children off from any contact with the universe. Consider, for instance, the trillion galaxies. We have learned so many awesome things about the galaxies of the universe, and yet how many of us have any direct experience of galaxies? Certainly some of us study galaxies in science class, but without a primal encounter with a galaxy, what good is such abstract knowledge?

In order to really learn about galaxies and about the birthplace of the universe, we need to struggle with this situation in which our lives and the lives of our children are almost completely encapsulated in human artifice. I am not now suggesting a return to former primitive life-styles, nor a rejection of technology, nor a romantic back-to-nature fantasy about abandoning the cities and living in communes. The first step is simple awareness. What we need is just the simple recognition that as we deprive ourselves and our children of direct contact with the numinous powers that fill the universe, we are choosing a diminished existence.

All of the industrial societies are captured, to varying degrees, by the idea frozen in Toronto's Skydome. Humans can become completely encased by this ingenious mechanical structure for weeks at a time. Once inside they never see the great sky or the blazing sun; they never smell the sweet Earth or touch the rough bark of the trees; they never hear the wind across the fields or the splash of a fish at twilight in the darkening lake. All of that and more is traded for a life in contact with cement, glass, astroturf, videodisplays, steel, and plastic. Take away the glitz and the crowds and you have the daily experience so many millions of our children wake up to each morning. Our first step can be just a suspicion that to trade in Sun and Moon and animals and galaxies for an ever-present television set may not be such a great deal after all.

If you've never had an experience of a galaxy, you have

before you one of the great opportunities in the universe. All that is needed is to journey forty or fifty miles to get out from under the urban pollution. If you live in the country, even better, for you are already there. Go with some friends, take a couple of blankets, a thermos or two, and as soon as it's nighttime lie on your back and behold that path of milky light that runs from one horizon clear through the great vault of heavens all the way down to the opposite side of the world. This is the Milky Way Galaxy, our home.

In an ultimate sense we know so little about the universe, so little about the dark past, so little about the future that is to come. But we do know that as we wonder over this great milky path surrounding us in the night, we are entering into an experience that billions of humans back through time shared. From every place on the planet, and from every period of history, humans have entered awe while contemplating this great light that encircles the world.

The beginning and the end of human existence is awe. We have that in common with all those countless humans before us who became mesmerized by the night's beauty. Our ancestors had their ways of explaining to themselves what it all meant. And now we too have a way, a new way, of explaining what is there. But in all cases awe is primordial. And if the explanation is a good one the awe will only deepen. The primary purpose of our explanations is not to remove the awe but to bring us into a more intimate understanding of these mysteries enveloping us and thus carrying us still deeper into the ocean of beauty in which we find ourselves.

During the modern period, we became so puffed up with our sense of superiority concerning science's explanations of the universe we did not even notice that we stopped wondering over the stars. It seemed reasonable to stop paying attention to the cosmos. Why wonder about something when we were convinced it was only a machine? Why pay attention to something when we had the mathematics that explained it? We regarded a scientific explanation as something that removed the mystery, so that in time we tricked ourselves into thinking that the mathematical explanations of phenomena were more significant than the phenomena! Believe that and before long you're living in the Skydome.

The beginning and the end is a primordial encounter with the great abyss of beauty that we call the universe. Not to enter such moments of awe, not to wonder over such majesty, not to live each day – at least for a moment or two! – floating inside a colossal and intimate mystery, is to live a life that is deprived. Even more, it is to live a life that is vulnerable to fundamental distortions.

As we lie on our backs mesmerized by the great Milky Way above us, we can ponder our recent discoveries in order to deepen this encounter. None of the billions of humans preceding us could experience the Milky Way as we can today. For when we contemplate the trail of milky light stretching across the sky we know this comes from the three hundred billion stars of the Milky Way Galaxy. We see these stars as a soft band of light because of the large-scale shape of our galaxy.

In its three-dimensional form, the Milky Way is sometimes compared to a pancake that has a bulge at the center. Another image sometimes used is that of an egg in a frying pan, the yoke as the bulge and the egg white as the larger and thinner sections emanating out from the hub. But I like to think of the Milky Way Galaxy as a gigantic manta ray. A manta ray is a fish with a flattened body, a slight bulge at the center, and two great dark wings for propelling itself through the oceans. The principal advantage in using the manta ray as a metaphor for seeing the Milky Way Galaxy is the ease with which the manta ray reminds us that the Milky Way is not sitting on anything; the Milky Way is gliding effortlessly through the dark universe just as the manta ray soars weightlessly through the seas.

Earth is two-thirds the way out from the thick center, and is inside one of the long thin wings. Since we are swirling around inside this galactic manta ray, at night we see stars in whatever direction we look, but when our line of sight is straight across the plane of the Milky Way's body we see many more stars, so many that their light commingles into that shimmering path overhead. This milky path continues all around the Earth. We cannot see it during the daylight because of the overpowering presence of the Sun. But as you lie on your back, imagining Earth within the body of the galaxy and the milky path stretching from one horizon clear across the sky to the opposite horizon and beyond, you can begin to experience the Earth as floating within this great body, the Milky Way, which itself is floating in the great body of the universe.

We now need to experiment with altering basic patterns of consciousness that have been set down in the primate line for seventy million years at least. As you lie on your back viewing the Milky Way, notice that an implicit assumption framing the experience is that of "looking up at the stars." This unconscious attitude is biologically rooted, for primates have had the up-down direction carefully coded into their organismic functioning. A chimpanzee awakening in the night in its nest in the trees needs to be instantly oriented by the up-down axis. Put another way, all the ancestral chimpanzees that needed more than a couple of seconds to figure out which way was up have long since met their demise by falling down.

Further strengthening our sense of up-down is the cultural coding worked out over millennia in terms of which the stars and heaven and God are "up," and the Earth is "below." Uncountably many generations of humans regarded Earth as the fixed place at the center of the universe above which the heavens turned. Such world-views simply articulate the implicit biological orientation of all primates. If an orangutan could speak, it too would regard the stars as far above, up in the sky; and if it were lying on its back on a field of grass at night, it too would think it was looking up at the stars.

So what do we do now, when our knowledge transcends the genetically and culturally coded assumptions concerning "up"? How do we orient ourselves in the universe now that we know a round Earth is sailing around the Sun, and that the Sun is a star similar to the other three hundred billion stars of the Milky Way, and that the idea of "up," which we might experience on Earth, has nothing to do with the dynamics of the galaxy as it spins its stars in their great orbits?

As you lie on your back beholding the Milky Way see if you can imaginatively free yourself from seventy million years of conditioning regarding our place in the universe. Such imaginative work is one of the great joys of being human and is at the core of all great achievements in art, science, or civilization as a whole. Imagine the Earth floating in space, and instead of picturing your own place on the

In order to enter the new story of the expanding universe we need to experiment with altering basic patterns of consciousness that have been set down in the primate line for seventy million years at least.

"top" part of Earth, arrange the picture in your mind so that you are on the "bottom" of this visualized Earth.

Now, as you lie there, imagine yourself peering down into the great chasm of the night sky. If your imagination is strong enough you can enter quickly into a new experience. Otherwise it might take some time, but the moment will come, in a rapid reorganization of phenomena, when all those stars will be experienced as down below, far, far below, and the amazing feeling accompanying this experience is a sense of surprise that you are not falling down there to join them. But of course you don't fall. You hover in space, gazing down into the vault of stars, suspended there in your bond with Earth.

Earth's gravitational power holds you, and you feel the strength of this bond in the pressure felt in your shoulders and along your back and buttocks and legs. We normally think of this pressure as coming from our "weight" but, in a strictly scientific sense, no being has any intrinsic weight. Rather, all bodies are capable of entering into gravitational interactions; and for us, the dominant gravitational interaction we experience is with Earth. It is not some intrinsic weight that keeps us here, for if the gravitational power of Earth and Sun were suddenly to vanish we would, with ever increasing velocity, soar away in a great rush down into that dark chasm of stars below us. It's not weight. It's the Earth's hold that keeps us suspended above the stars.

As you lie there feeling yourself hovering within this gravitational bond while peering down at the billions of stars drifting in the infinite chasm of space, you will have entered an experience of the universe that is not just human and not just biological. You will have entered a relationship from a galactic perspective, becoming for a moment a part of the Milky Way Galaxy experiencing what it's like to be the Milky Way Galaxy.

The path of white that we are dreamily contemplating has the power of the gods. The destiny of our Earth and our Sun and the planets and our own bodies is controlled by that Milky Way sweeping through the night sky. That milky white path is whipping our entire solar system around the outer edge of the galaxy at a speed of 180 miles a second. As you lie there and count to twenty, all the animals and forests and the entire Earth and Jupiter and the asteroids and even the great Sun are flung a distance equal to the width of the North American continent.

I sometimes wonder over the fact that if I could lift a ton of bricks and toss them a few feet I would be on every TV screen in the world. But the Earth is a billion trillion times more massive than a ton of bricks, and the Sun is a million times huger than Earth. And all of that and more is not just lifted up and tossed a few feet but has been flung 180 miles each second of the day, all day long, all year long, for five billion years now. The origin of such titanic power is that milky white path that we contemplate as we are held suspended over its immense powers. Why would people watch me lift a ton of bricks when they could behold each night a dynamism a quintillion times mightier?

We modern peoples have said some very questionable things about what a great advance scientific humans represent. To fall into the trite conceptions about how limited the earlier primal peoples were is an easy mistake for a consumer to make – I mean, after all, they had no dishwashers, Chevrolets, or Apple computers. But they did understand something central that escapes us modern and post-modern peoples. We have forgotten what it was and, even worse, we've forgotten that we've forgotten. In their work of orienting themselves within the great determining powers of the universe they achieved a peace that we can only guess at.

They had little technology, they had no electrical appliances, they were vulnerable in a thousand ways, but they lived in a form of consciousness far beyond what we normally experience in our daily lives within our industrial cages. With each morning they awoke to a numinous universe. With each morning they awoke to a reality we usually experience only in our dreams where we are swept up into a great adventure involving the entire universe. If we do not know what it was that they experienced we can hope that as we develop our own cosmological relationships with the powers of the universe, we too will wake up one morning in an enchanted world in which we have a role to play and are able to speak to our children about the ultimate things.

7

The Large-Scale Structure of Space and Time

Our principal task as humans is to live in the universe. In a superficial sense of course everyone lives in the universe because we're all physically here. But in an intellectual or spiritual or emotional sense, most of us live elsewhere. This is indeed a strange situation, but it is a deformation that humans have succumbed to over and again throughout history.

Sophocles made a study of this interesting possibility of a split life in his drama *Oedipus Rex*. This is the story of a man who arrives in a kingdom one day and through an unfortunate string of events ends up killing the king and marrying the queen, completely ignorant of the terrible truth that the king is his father and the queen his mother.

We can say that even though Oedipus was made king of Thebes, he did not actually live in the kingdom of Thebes. For to live in the kingdom means to live in proper relationship with the members of the kingdom. But Oedipus was in proper relationship with no one: he was the husband of his mother; he was the murderer of his father. And yet if anyone had asked Oedipus whether or not he lived in the kingdom of Thebes he would think the questioner insane, for where else could he be? But in the deeper sense of his own understanding of his essential nature and role in Thebes, he was not a member of the kingdom; he was an abomination of the kingdom.

We too regard ourselves as living on Earth. But we do not live on Earth in the sense of living as members of Earth's Community. Both in our activities, as well as in our own understanding of ourselves and Earth, we are simply not members of Earth's life. We live in this split condition, thinking we are members of Earth, unaware that we are the destroyers of Earth.

An easy way to touch the ragged edges of this gaping hole in our consciousness is to take the "local universe test." You ought to try it sometime. It's easy to do. You simply invite someone to visit you who lives at least twenty miles away and who has never visited you before. You can give verbal instructions on how to get to your abode over the telephone, but the one rule is this: in your directions you may refer to anything but human artifice.

You may refer to hills, oak trees, the constellations of the night sky, the lakes or ocean shores or caves, the positions of the planets or any ponds, trails, or prairies, the Sun and Moon, cliffs, plateaus, waterfalls, hillocks, estuaries, bluffs, woodlands, inlets, forests, creeks, swamps, bayous, groves, and so on. Whenever your friend gets stuck, she is free to phone you for more directions, but the rule for her is that she must describe her location without referring to any human artifice. Do this a couple times and you will learn directly how we, like Oedipus, do not really live within the Earth Community.

So where do we live? In a physical sense, we live in industrial artifacts designed to keep us inside and the universe outside. It is for this reason that the primary concerns of our hearts and minds are overwhelmingly focused on the requirements imposed upon us by life in these industrial artifacts. The resulting American mind of the late twentieth century is a strange brew. We have some good things floating around in there, but they're mixed in with some of the most soul-shriveling trivialities. It's depressing to realize how easily any one of us can talk endlessly about matters of utterly no significance, yet almost all of us become walls of cement when asked the simplest things concerning our actual lives in the universe – such as, "What species of life live in my backyard?" Our eyes and hearts are so crammed full with the many demands of consumerism that we rarely notice even the most basic contours of the life place in which we live.

Our children can reproduce dozens of tinny tunes from advertisements, and yet they cannot distinguish between the songs of the meadowlark and the mockingbird. Our preschoolers, even before they can speak, can recognize the corporate images of so many of our commercial enterprises, and yet few if any of them can draw the insects or trees or flowers or mammals of our life place. The truly great cosmic event each year of their lives is the rhythmic sequence with which spring explodes into being, but the chance of finding a single child, just one, who knows and celebrates this awesome rebirth of life is much smaller than the likelihood of walking into the executive suites of Nintendo or Sega Genesis and convincing them to stop dumping the relentlessly repulsive violence of their videogames into our children's lives.

Here at the tail end of the millennium we take our steps into the future one at a time. We can't, with a single command, halt the destruction of the soils and the animals and the children, but we can with a single decision begin the search for ways to align our own energies with the creative, restorative, and healing directions already taking place. In terms of our children, we can't, with a snap of our fingers, transform their education from industrial to Ecozoic, where by "Ecozoic" I mean a form of education that would initiate our children into the ways of the universe. We can't instantaneously wash out all the trivialities and toxins and provide instead the knowledge and information essential for their planetary future. But we can begin to make a difference. We can begin by introducing our children and ourselves to the universe. We can start by showing them they are part of a Big Picture; they have a place and a role in this enveloping activity. In time, if they are fortunate, they will eventually learn to regard all the things of the world, even the briefest breath of the tiniest gnat, as woven into a single, comprehensive, coherent whole.

Here's a way to make a beginning. Get someone to show you where the constellation Sagittarius is, and then take a child outside and direct your attention there. When you do so, you will be gazing into the very center of the Milky Way Galaxy, the hub around which all three hundred billion stars revolve. By introducing young people to our galaxy as a whole and by allowing them the opportunity to understand themselves as living beings within this enveloping galactic process, they begin the deep journey into a much vaster context in which they can find the true meaning of their lives. Knowing the sacred direction toward the center of the galaxy and returning to it over and again will be part of the empowering process that will enable them, slowly and subconsciously, to think of themselves not just as political or economic entities. They will learn that they are, primarily, cosmological events.

To speak in terms of light, we are just under thirty thousand light-years from the center of our Milky Way Galaxy. A light-year is a measurement of distance. It is the distance light will travel in a year's time, around six trillion miles. For instance, light from the center of the galaxy traveling 186,000 miles each second reaches us today after having streaked toward us for the last thirty thousand years. So if tonight you direct your gaze toward Sagittarius, some of the photons of light that reach you left the galactic center when giant woolly mammoths roamed the North American continent.

The photons were created in those long-ago moments when the paleo-Indians revered and hunted these giant beasts. The saber-toothed tigers, like the Indians, stalked the mammoths and would continue to hunt another twenty thousand years before disappearing in the amazing event of extinction. And in every instant of the saber-toothed tiger's existence, as generation after generation made its way hunting and mating and sleeping and stalking, in each instant the photons from the center of the galaxy soared silently through space on their journey toward us.

When a child looks toward the galactic center, she needs to remember the woolly mammoths and the saber-toothed tigers and the paleo-Indians. And the brief time in which civilization exploded onto the scene. All that great volume of time was necessary for the photons rushing at light speed to reach us tonight as we gaze into our galaxy's center.

Unless we live our lives with at least some cosmological awareness, we risk collapsing into tiny worlds. For we can be fooled into thinking that our lives are passed in political entities, such as a state or a nation; or that the bottom-line concerns in life have to do with economic realities of consumer life-styles. In truth, we live in the midst of immensities, and we are intrinsically woven into a great cosmic drama. Economic and political concerns are of real importance, but children need to understand that whatever importance and value these concerns have derive ultimately from our encompassing matrix and its deepest meanings. To be out of touch with this cosmological context is to risk living within a shrunken and distorted version of reality, such as happened with Oedipus.

Besides the direction toward the center of the Milky Way Galaxy there is one more orientation we can give to our children, and that is Andromeda Galaxy. If you direct the children's gaze toward the constellation also called Andromeda they will see there a faint blur of light unlike that of any of the stars. With an average pair of binoculars you can even see a spiral structure in the blur of light. This is the outer horizon of the naked human eye. This is Andromeda Galaxy, slightly larger than our own Milky Way, coming to our eyes from 2.5 million light-years away.

Something happens deep in the soul of a child gazing at a blur of light while knowing this blur is a galaxy with hundreds of billions of stars. Something happens inside, a spaciousness opens up in psychic congruence with galactic spaciousness. That blur of light required millions of years to make it to this young human eye, and it is now an eye within a being who knows all this. Aware of the larger dimensions of this experience, her inner world opens out as the light enters in.

On Earth 2.5 million years ago humans were first discovering the use of tools, thereby entering into the long drama of seizing hold of the universe in a new way. On the day that these first humans were shaping the first human tools light as always exploded away from Andromeda, but this particular light had an extraordinary destiny; for as it roared away from Andromeda and rushed toward the Milky Way these inventive humans continued to develop their tools, their minds, their sensitivities, and their understanding, until at last, 2.5 million years later, they had discovered their place in the space and time of the universe, just in time to take their children out into the night in order to capture and wonder over those very photons of light that had left so long ago. Certainly to gaze at Andromeda Galaxy is to enter galactic spaciousness, but it is also to enter into a new sense of time concerning the human venture. The human journey is as immense in time as the galaxies are distant in space.

Andromeda and Milky Way are large galaxies, and they slowly pinwheel about each other. Each has a dozen or so galaxies encircling it. The best known satellite galaxies going around the Milky Way are the Magellanic Clouds, Fornax, Draco, and Sculptor. Poor Andromeda has only very dull names for its companion galaxies. For instance, M32. We have so little experience of them we have not yet been able to name them properly. Before the twentieth century we didn't even know there were other galaxies in the universe, so naturally it is going to take some time before we move beyond just assigning numbers to each galaxy.

The mega-system consisting of the Milky Way, Andromeda, and all of their encircling satellite galaxies is spread out over several million light-years, contains at least half a trillion stars, and is called by astronomers the "Local Group." There's something rather humorous and sad about that name, but we can consider it an interim tag soon to be replaced by the names the poets of the twenty-first century invent as they learn this new story of the universe.

Our Local Group of galaxies itself is but one satellite of a still vaster system. Just as planet Earth revolves around the Sun, so too does our Local Group revolve around a central hub called the Virgo Cluster. Virgo is a gigantic cluster of a thousand galaxies fifty-three million light-years away. Our Local Group along with hundreds of other galaxy clusters all revolve around the Virgo Cluster. When we reflect upon this whole supercluster system with its thousands upon thousands of galaxies, each of which might contain ten million intelligent planets, the primate mind begins to dissolve. Just recently our major challenge was to catch the next branch as we swung through the forests. Now we are left to ponder the immensities of this cluster of clusters, the Virgo Supercluster.

If we can now bring ourselves to imagine this immense supercluster as being a single white dot, then the universe as a whole consists of ten million of these that are floating, drifting, and twirling as apple blossoms do when in the early spring a gust of wind frees them from their branches and carries them aloft into the blue sky.

8

The Story Came to Us

There is no theoretical or scientific reason that requires the birthplace of the universe to be fifteen billion light-years away. On the contrary, it is an empirical discovery; it could have been found to be otherwise. The birthplace was located through observation. Our distance from our cosmic origin was measured, and fifteen billion light-years is what it turned out to be. It's not the exactness of the number that is the point here. In fact the exact distance is some number between ten and twenty billion light-years. Our data do not allow us a more careful determination yet. The point here is that these numbers do not follow from some prior scientific or philosophical theory about the universe; they follow from an extremely sophisticated measurement of the universe.

When the astronomers Vesto Slipher, Edwin Hubble, and others first began gathering data on the motion of galaxies, they had no idea they had embarked on an empirical investigation that would lead to the discovery of the universe's birthplace. Indeed, few considerations could have been further from their minds. They were simply observing the universe and carefully attending to the movements of the galaxies.

It is significant that the very people who discovered the birthplace were shocked by the discovery. Some were depressed by it. Rather than trumpet their great achievement, they reported it with a sense of bafflement. Hubble, the very scientist whose work was central for convincing the scientific community that the universe had a birthplace, refused to make any interpretations. He simply published what he found, however strange it might appear to him. And Einstein, whose theory provided the deepest interpretation of the data, began by actually altering some of his best ideas in order to avoid confronting their radical implications. At the very least we need to understand that they were not projecting their mathematical theories nor imposing their philosophical perspectives on the universe. They were, rather, attending to the movements and structures in the universe. They were, rather, listening deeply to the patterns in the photons that came to them each night in their observatories.

The seemingly innocent phrase "came to them" hides yet another important truth that we need to bear in mind. Edwin Hubble and other scientists are members of the life of Earth, revolving around the Sun, in the three hundred billion stars of the Milky Way Galaxy, which pinwheels through our Local Group, all of it spinning inside the Virgo Supercluster, which dangles as one of ten million in the great universe. Scientists, and everyone else as well, can experience only what the universe brings to us here. What we know about the universe is gotten by listening to and reflecting on the news the universe brings.

As an illustration we can return to our discussion of Andromeda Galaxy. Every night throughout all the millions of years of human existence new photons of light have arrived from Andromeda. But it was not until we had painstakingly developed all the necessary tools, including telescopes and mathematics, but more generally all the arts and languages and conceptual capacities of modern *Homo sapiens,* that we could penetrate into the information these light particles carried. The story of Andromeda and its three hundred billion stars and its circling spiral structure has been present to Earth all along. And no one ever had to journey to Andromeda to learn her story. Rather, humans had to develop the sensitivities necessary to awaken to the story that has been here throughout our entire existence.

We scientists tend to use phrases that speak of our ability to "reach deep into space." We speak of our invention of instruments that "probe the farthest regions of the universe." These are certainly valid expressions, but they can also give the mistaken impression that we actually "reach out" in some literal sense. There is no reaching out. Rather, we reach into our immediate experience.

We reach into a droplet of the universe, and we find there photons with wondrous stories from the farthest regions of the universe. All the books on the distant galaxies, all the volumes and journal articles on the large-scale structure of the universe, all the tomes on the dynamics of neutron stars, all the photographs of the brilliant nebulae, all the studies of super red giants are, strictly speaking, explications of the stories that exist in each cubic centimeter of the universe. These stories have been there for all of human history, but not until recently have we been able to read them. Optical telescopes, infrared detectors, x-ray diffraction, spectroscopic devices – all such instrumentation of modern astronomy aims at becoming sensitive to the news of the universe contained in each drop of the universe.

The discovery of the birthplace of the cosmos, then, is the discovery of the story that has been present from the beginning. The universe's primordial origin has been here with us for millions of years. It showered us from all directions as we wandered the African savannas and built our mud-brick abodes along the Nile. Our own generation is simply the one to emerge at the time when human consciousness has become subtle enough and complex enough to awaken to what the universe has been telling us from the beginning. If it is true that the discovery of the universe's birthplace is the work of a few spectacular geniuses, it is also true that the discovery is the work of the entire human venture. Every genius, even the greatest, comes out of and works within this developing awareness. The deepening of consciousness necessary to hear the story of the universe required the complexification of consciousness coming from all four million years of the human journey.

The human is the space created in the universe process for hearing and celebrating the stories of the universe that fill the universe.

9

Nighttime and Cosmic Rebirth

The more significant proposals humans make to one another often involve the light of the Moon. I wonder why this is so? After all, on a purely statistical basis, we live primarily in office buildings, traffic jams, dwellings, factories, and shopping malls.

It might have to do with courage. Sometimes it is hard to believe what we hear surging up from our hearts and we need to be half-hidden by the dark in order to blurt it out. Or perhaps we hear our deepest intelligence only in the dark. Perhaps night is a time when great discoveries are made because a beauty lives in the night that does not show itself at other times. Even though we in the industrial societies have forgotten this more than most people, the lesson of the night's sacred nature is so often forgotten. Immanuel Kant lamented our forgetting by claiming that, "If the stars came out once a century, humans would stay up all night long marveling at their beauty."

What about the day? The surprising thing is that the beauty is still there but hidden by another overpowering kind of beauty – the scattered blue light of day's sky. The stars shine always; they fill the skies both day and night. But night is a time free from daily scattered distractions. Night is a time when the presence of the stars can be more deeply felt, when the news of the universe can be more deeply attended to.

The ancient astronomers, the first cosmologists, and the shamanic storytellers often told their stories at night. The concerns of the day, however important they might seem in the sunlight, usually amount to nothing more than unwelcome distractions in the night when the great story is told in the glow from the fire's embers and Moon's journey through the branched shadows of the trees. It is in the peace that the night brings that something immense can stir in the depths of the listener, things not suspected during the day. Or if suspected briefly, then so quickly forgotten as some daytime urgency forces out the haunting music. Late, very late, after the Sun is gone – such is the time for the great surprises deep in the listener's soul. Such is the time to ponder the mysteries of one's existence. For what was invisible as we dashed about from one errand to another suddenly stands out, magically present, no longer willing to be ignored.

As with so many epochal transformations of human consciousness, the discovery of the center of the universe took place at night. This discovery required the intelligent development of all two million years of human existence, a development that ended in a profound sensitivity to the news the universe brought to Earth. This sensitivity was not located in just one particular human but was present in a small number of humans, including Hubble and Slipher, as well as Carl Wirtz, Howard Robertson, William de Sitter, and a few more. But in a symbolic sense I want to speak of the essential news as breaking forth in one particular human, as surfacing

The first cosmologists and shamanic storytellers often told their stories at night – the time for the great surprises deep in the listener's soul.

for the first time within our century's most celebrated scientist. To tell the story of the discovery of the birthplace of the universe is to tell the story of Albert Einstein.

The crucial period was the second decade of the twentieth century in Berlin. And although a novelist or a historian could perhaps detail in a straightforward manner the events that took place in that history-making time in Einstein's life, a cosmologist has a particular difficulty. My aim is to tell the story of the universe and, in speaking of Einstein, I am interested in him as one particular story within the encompassing story. So in order to tell the story of Einstein I must first tell the story of the universe as a whole and then show how the story of Einstein fits into the encompassing story.

But here is the strange thing. How can I tell the story of the universe without first speaking of Einstein, for he is a fountainhead of the story of the universe as a whole?

Einstein can be seen as a power that destroys one world and creates another. In approaching Einstein bent over his desk at night through the years of World War I we see one event if we use the pre-Einsteinian physics of the modern period. But we see a larger and more inclusive Einstein-event if we use the physics that flowed from his pencil one particular night. He was, at that moment, literally destroying and then recasting a central code of the Newtonian cosmology. In order, then, to understand what was happening there in the upper floor of the Berlin apartment we need to understand the event through the very insights unfurling on the page.

No one in the entire world at that time could understand the reality of the Einstein-event. How could they? All had minds structured by one or more of the pre-Einsteinian cosmologies. In the West and in industrial countries generally this would include the Newtonian world-view, a brilliant and profound system of insights that had endured for three centuries, all the while guiding scientific research and offering an authority to the social and political and economic structures congruent with that world of meaning.

Within the Newtonian framework the Einstein-event would be described along the following lines: "Albert Einstein, on November 22, 1914, articulated the gravitational dynamics of the universe in the form of his field equations." Or I could put it this way: "On November 22, 1914, Albert Einstein created his General Theory of Relativity, extending his Special Theory of Relativity into the case where gravitational effects are taken into account." These sentences are true but they miss in the grossest way the real drama taking place that night. To get a sense of what was happening we need to reflect on Einstein as he sat at his desk stunned by what had appeared on the white paper.

To everyone alive at that time, his equation would appear to be just some letters and a few numbers scrawled onto the page. But to Einstein they revealed something unbelievable about the universe as a whole. Even though by this time he must have grown somewhat accustomed to the regularity with which shattering truths dropped from his fingertips, Einstein was stunned into bafflement by what he was seeing. Through these symbols the universe whispered that it was expanding in all directions. No one in three centuries of modern scientific work had imagined such a possibility. All his life Einstein had assumed the universe was an unchanging infinite space. Now he was confronted with the idea that space was expanding in every direction. This was not a minor modification. This was an idea that, if true, would shatter the world-view of everyone, Einstein included.

Newton, Herschel, Galileo, Kepler, Darwin, Curie, and all scientists of the modern period took the universe to be an unchanging macrocosm. The universe was perceived as a vast and fixed place, a celestial container that housed the stars and planets and everything else. Then some strange symbols appeared, equations on the back of an envelope, mocking our former perspectives. A hieroglyphic arcanum in the language of differential geometry contained a bizarre truth that scientists would be unraveling for centuries: the universe had erupted as a single ultimate density of being fifteen billion years ago. The universe had a birthplace, a center in space, the universe had an edge to its existence, the beginning point of time.

Can anyone wonder at the fact that Einstein rejected this truth? That he lost his nerve? That he altered the equations to hide their difficult truth? In all this he was only too human, for how many of us are capable of accepting, all at once, the full truth when it comes in the form of a knife? How many of us let it cut through our hold on a false version of reality without first administering some edge-blunting that allows us to cling to one or two accustomed fictions?

Einstein doctored his equations. He added a mathematical term now known as the "cosmological constant." By altering the equations he took away their secret story of expansion, and thereby preserved his attachment to an unchanging universe. It was only these altered equations that he allowed out of his study. Perhaps he was confident no one would notice. Perhaps he was hoping everyone would view the equations as another straightforward contribution to the scientific canon. But, as in those mythological stories where to change one word of a sacred chant throws the entire universe to its ruin, so too with Einstein's sacred code.

Science is a collective enterprise. No genius, however magnificent, is sufficient alone to speak the truth. In Russia the mathematical cosmologist Alexander Friedmann, when pondering Einstein's equations, made what for him was a startling discovery. If one worked with these Einstein equations just a bit, dropping off this one strange term, why, it turned out that Einstein's equations pointed to an expanding universe!

Imagine Friedmann's excitement as he rushed to write a letter to Einstein to tell him of his discovery. I only wish someone had snapped a photograph of Einstein's face when he unfolded the letter from this excited young mathematician. Einstein thought he had put that business behind him, and now it was right back in his face. C. G. Jung's melancholy summation comes to mind: "One finds one's destiny on the path one takes to avoid it."

As it turned out, even Friedmann's work failed to dislodge Einstein from his error. But in the next decade Edwin Hubble in California trained his telescope on the distant galaxies and saw that they were in fact expanding away from us. Finally, in the 1920s, when Hubble invited Einstein down to Mount Palomar to see for himself, Einstein became convinced that the old idea of the universe as a fixed, unchanging macrocosm, the old idea that the universe was simply a giant box, was wrong. Only when Einstein saw with his own eyes the galaxies expanding away from us did he realize that his original insight concerning a dynamic expanding universe was in fact the truth.

The universe was expanding. The universe had an edge to its existence. The universe had a birthplace. All of these had been predicted by his original equations. At the end of his life, when Einstein reflected on his journey, with its historic work at the highest level of genius through five decades of the twentieth century, he concluded that the doctoring of his field equations in order to escape their prediction of an expanding universe represented "the greatest blunder of my scientific career."

10

The Place Where the Universe Began

If we look out beyond the Milky Way Galaxy and out beyond the Local Group of galaxies, if we look out into the great sea of galaxy clusters, we find that the clusters are moving away from us. This is true whether we look at the clusters that are straight above the Milky Way Galaxy, or to the west, to the south, to the east, or to the north, or down below the Milky Way. In every direction we look we find the clusters of galaxies expanding away from us.

Reflecting on this bizarre discovery, which was predicted by Einstein's original field equations and confirmed by Hubble's galactic measurements, George Gamow and his collaborators concluded that the universe as a whole must have begun as a point that exploded in a titanic burst of energy. For Hubble discovered something more, something that made this conclusion inescapable. Not only are the galaxies expanding away from each other, they are moving away with a velocity related to the space between them. *The farther apart they are, the faster they are flying away from each other*. More precisely: galaxies twice as far apart are sailing away from each other twice as fast. Galaxies ten times as far apart sail away ten times as fast.

The mathematical conclusion could not be more startling: by tracing the trajectories of the superclusters of galaxies backward, we find an event of cataclysmic energy where all trillion galaxies are brought into a single ineffable point, the birthplace of the universe, the initial singularity of space-time, the center of the universe. There, in that place, the entire cosmos began as a pinprick, a sextillion-ton pinprick layered with the power to thunder forth into the beauty of existence.

This vast and complex tapestry of being – the Magellanic Clouds, the Milky Way, the Andromeda Galaxy with its local satellite galaxies, the thousand galaxies of the Virgo Cluster, and all the shining superclusters lining the Great Wall – all these are sailing off from an initial birthplace. And each supercluster rushes away from every other supercluster, all of them emerging from a single point.

The idea that the universe began in one place is certainly an ancient one in human history. The image of a birthplace of the universe occurs in the mythical and classical forms of consciousness, and possibly even earlier. Such images as "the cosmic egg" that cracks open and gives birth to all phenomena are found in Neolithic cultures around the planet. Thus, the scientific discovery has this dual nature of being both old and new simultaneously. The mathematical theories and observational data are indeed new and represent a new kind of knowing; yet the overarching images are very old.

We need continually to distinguish the scientific enterprise from earlier forms of inquiry in order to avoid the two most common errors: insisting that scientific understanding is altogether divorced from other kinds of knowing, or claiming that in essence there is no difference between the modern scientific and the other forms of knowledge. Each mode is primordial; each is qualitatively distinct from the others. Science is an investigation of the universe rooted in empirically verifiable physical detail and is complementary to our earlier and more intuitive investigations of reality. The aim is not to eliminate one way of knowing in favor of another; the aim in an ultimate sense is an integral understanding of the universe grounded in both the scientific empirical detail and in our primordial poetic visions of the cosmos.

One of the principal differences in these approaches is the scientific hypothesis. Science does not proceed from one well-established fact to the next. Science proceeds rather with a series of hypotheses that can be tested and rejected on the basis of empirical discoveries. Science then offers a hypothetical story of reality that is constantly revised, but stronger and more dependable with each generation for the simple reason that an ever larger body of experience is appropriated into its intellectual framework.

We can consider the process of how science establishes knowledge by looking at Gamow's postulate that the universe began in a great explosion. When Gamow initially made his postulate, it became just one among several different theories about the universe and its development. Each of these accounted for the facts of the universe that we knew, but each also made predictions about the facts of the universe we would discover in the future. Gamow's postulate of a fiery beginning slowly advanced in status from postulate to scientific knowledge when, over the course of several decades, the predictions following from his theory turned out to be true, one after another.

To focus on just one example we can consider Gamow's prediction that there must be, throughout the universe, a form of radiation that came from the primordial birth of the universe. This and other predictions were debated by scientists, philosophers, and theologians without anyone establishing whether or not they represented actual knowledge. In order to know if the theory provided knowledge someone had to check with the universe. Thus it was that in the 1960s all theoretical debate was swept aside when Arno Penzias and Robert Wilson actually detected the cosmic background radiation predicted by Gamow. Here was the dim glow left over from the eruption of the universe at the beginning of time. Penzias and Wilson captured the photons – the particles of light – that had been set in motion fifteen billion years ago when the universe erupted into existence.

With this discovery in mind we are ready for another trajectory into the meaning of the sentence "The birthplace of the universe is fifteen billion light-years away."

The light Penzias and Wilson discovered originated in the primeval fireball itself and had been traveling for fifteen billion years. To trace these photons back to their origin fifteen billion light-years away is then to follow the trail back to that sacred place where the universe first flared forth into existence. That place from which the photons left is the cosmic center, the world's navel, the sacred origin point of being. That is the place endowed with the stupendous fecundity necessary to give birth to the cosmos.

Whereas former mythical consciousness might create images of our cosmic birth, and whereas classical consciousness might reason philosophically and theologically about the universe's beginning, scientific consciousness located in an empirical way the physical birthplace of the universe. I neither want to claim that scientists were the first to discover the story – obviously humans had for millennia spoken in various languages about the birth of the universe – nor do I want to pretend that science is only corroborating what earlier peoples already knew – we now have some details of the story that were never even suspected earlier. Instead, I want to remember that humans have a variety of paths to the truth and when these various routes arrive at a common consensual knowledge we have the possibility of a story of the universe that can guide us as a whole species as we enter a new millennium.

If the news from science consisted solely of this fact that the birthplace of the universe could be gotten to by tracing the path of the primordial photons back to their birthplace, we would already have a magnificent new understanding of the universe, for now we would have measured the place of the cosmic egg's appearance. But the story has one added complexity. In order to achieve a fuller understanding, we need to move beyond a paradox that I have avoided mentioning explicitly but that is most likely hunkered down in the basement of your mind, with its apparent irrationality sapping away your psychic energy.

11

A Multiplicity of Centers

I stated earlier that when Edwin Hubble began watching the galaxies, he discovered that the galaxy clusters were moving away from us in all directions. What I did not point out then is the startling conclusion that, in terms of cosmic expansion, *we find ourselves at the center of the cosmos*. This is indeed a strange and most unexpected development.

If such a discovery had been made during the medieval period of Europe it would have caused no surprise at all, for the presiding world-view at that time put the Earth at the center of the universe. To extend Earth's centrality to the center of expansion of all the galaxy clusters would have made no great demands on the medieval mind.

But, of course, when Hubble made his discovery we no longer lived within the medieval world-view. We had already learned from Copernicus that Earth is not a fixed point at the center of the universe. Earth is one planet moving around one star, which itself is one of the three hundred billion stars of the Milky Way Galaxy, which in turn is one of a trillion galaxies in the wide universe. If we have learned anything over the last four hundred years since Copernicus initiated this great search, we have certainly learned that the Earth is not a fixed center around which all the planets and stars revolved.

And yet, here was this new revelation placing us at the center of the cosmic expansion of the galaxy clusters. Our Virgo Supercluster was not moving at all, and all of the other superclusters were moving away from us. What were we supposed to make of it?

Complicating our challenge is the consternation of the scientists who discovered the expansion. For if we could convince ourselves that it was Einstein's secret hope to put us at the center of the universe, we could dismiss his and similar work as an ideological imposition upon the universe. On the contrary, as we have already examined explicitly in Einstein's case and as was true of other scientists as well, the very discoverers of the cosmic expansion were repulsed by the idea, and they did everything they could to avoid accepting it.

Had the cultural and personal biases of these scientists determined what they saw, they would not have found all the clusters of galaxies moving away from us so symmetrically that we were placed at the very center of this cosmic expansion. If they had been free to distort the data to fit their own preconceived notions about the large-scale nature of the universe, they would have announced that all the galaxies were fixed with respect to each other, which is what Einstein's doctored equations suggested. Or, if they couldn't have an unchanging universe, they might have preferred one in which all the galaxies were moving in the same direction, a Great River of galaxies. Then at least we would not be in any special place. Such a discovery would then fit smoothly into modern culture, for it would suggest we were

When we picture the cosmic birth as some kind of explosion taking place away from where we are observing, just where are we standing? How can we stand outside the universe if from the beginning we are woven into this birth?

insignificant, without cosmic meaning, just as Friedrich Nietzsche and Jean-Paul Sartre and Bertrand Russell and so many other modern philosophers taught. Earth would be just one bit of bark swept along with the Great Current.

What Hubble discovered did not fit our modern culture's preconceptions but in fact disrupted them. But instead of altering the data he published it. He offered no philosophical justification for or against the data. He simply made public what he had discovered there at the cutting edge of human awareness: in terms of the universe as a whole and its fifteen-billion-year expansion, we happen to find ourselves at the very center.

Hubble's discovery is not a contradiction of what Copernicus learned; Hubble's discovery is instead the completion of the cosmological exploration that Copernicus started. Copernicus initiated an investigation that removed the Earth from the center of the universe then removed the Sun from the center and then removed everything from the center. But after four hundred years of empirical inquiry, a great reversal has taken place, one that shows us the center in a universe vastly huger than the solar system and nearby stars Copernicus and Galileo were aware of. We do not return to the cozy medieval geocentric world but enter an immense evolutionary cosmos, a cosmos that is centered on its own expansion. In order to fully appreciate this new understanding of the cosmic center we must now deal with the seeming paradox at the heart of the data.

I've presented two discoveries that seem in conflict with each other. First, in terms of the light from the beginning of time, which was first detected by Penzias and Wilson, the birthplace of the universe is fifteen billion light-years away from us. Second, in terms of the expansion of the galaxies, which is Hubble's discovery, we are at the very center of the universe. We need to consider this strange situation where we are simultaneously at the center of the cosmic expansion and fifteen billion light-years away from the origin of the cosmic explosion. The paradox is this: How can we be both at the center and fifteen billion light-years away from the center?

We have such difficulties with this discovery because our minds have been shaped and educated in a culture firmly rooted in the Newtonian world-view. Even though we now know Newtonian physics is not adequate for complete understanding of the vast evolutionary universe that was discovered since Newton's death, we are nevertheless stuck with Newtonian consciousness because it forms the foundation of our major institutions – including our educational systems. The challenge of understanding an Einsteinian universe is a real challenge indeed. We need to reinvent our very minds so that we do not distort the discoveries by holding them in Newtonian categories that are unable to touch the truth we have discovered.

To give a single example that bears on our discussion of the center: When we hear that the universe began in a great explosion fifteen billion years ago, we picture this as something like a Fourth of July fireworks explosion. First there is just empty space, then there's this great explosion of colors in all directions. We are forced into picturing the birth of the universe in this way because Newtonian cosmology regarded the universe as a giant fixed space inside of which things move about and gather together and so forth. The shaping of our minds in childhood already compels us to picture the birth of the universe as an explosion taking place in an already existing space.

But this understanding of the universe's beginning is both false and utterly misleading. The birth of the universe means not only the birth of all the elementary particles of the universe and not only the birth of all the light and energy of the universe, it also means *the birth of the space and time of the universe.* There is no pre-existing Newtonian space into which the universe explodes forth. There is no external Newtonian timepiece clicking away outside the universe. Space and time erupt together with mass and energy in the primordial mystery of the universe's flaring forth.

The simplest way in which to see the inadequacy of our mind's Newtonian assumptions concerning the universe's beginning is to ask a simple question. When I picture the cosmic birth as some kind of explosion that is taking place off in the distance, away from me, away from where I am observing it, just where am I standing? What provides the platform for my feet? How is it that I can stand outside the universe and watch its birth if I myself, from the beginning, am woven into this birth?

A re-education of the mind is necessary to make sense of what we have discovered. The central archetypal pattern for understanding the nature of the universe's birth and development is omnicentricity. The large-scale structure of the universe is qualitatively more complex either than the geocentric picture of medieval cultures or the fixed Newtonian space of modern culture. For we have discovered an omnicentric evolutionary universe, a developing reality which from the beginning is centered upon itself at each place of its existence. In this universe of ours to be in existence is to be at the cosmic center of the complexifying whole.

If there are Hubble-like beings in the Hercules Cluster of galaxies seven hundred million light-years away, and such creatures are pondering the universe from that perspective, they will also discover that the galaxies in the universe are moving away from them. They will thus conclude on the basis of this evidence that they are at the center of the universe's expansion, and they will be correct. Our Newtonian minds might experience discomfort in the task of appropriating this knowledge, but our personal difficulties do not change the nature of this universe. Just as Einstein's first reaction when he was given a glimpse of our omnicentric evolutionary cosmos was to pull back and insist the universe could not be like that, so too in our own struggle we sometimes wish that the universe were not so complex, not so mysterious. But the universe will be what it will be regardless of whether or not we humans accept it as it is.

There is one image in the scientific literature that can give some assistance in the journey into an omnicentric universe. I offer it with some misgivings because the image, however helpful in some ways, is also inadequate in others, as I will point out. My hope is that it can help us take a first step out of the false view of the universe as a fixed space. And perhaps it might help awaken in others more adequate images of the nature of the universe that will become a regular part of our cosmological education in the future.

Imagine you are inside a loaf of raisin bread as it is being baked. The crucial point is to begin your imaginal work from *within* the process rather than outside of it. So we have to forget the nagging Newtonian questions concerning the oven, or the loaf's crust, or any other concern that arises when we attempt to understand the process from *outside.* We are inside a cosmic process; even our thoughts about this process are simply yet another interesting current of micro-events taking place inside the great macro-event of the fifteen-billion-year development.

So in this particular raisin bread image we need to focus our imaginations on being in the very midst of the baking raisin bread. In particular, imagine yourself on a raisin and just look around. You will see that all the other raisins are moving away from you as the bread bakes, so that in terms of the bread's expansion you find yourself at the very center. And anyone else on any other raisin throughout the loaf would come to a similar conclusion – hence, we have in this raisin loaf a model for an omnicentric reality.

But there's more. Suppose you now try to determine whether or not you and your raisin are moving with respect to the bread itself. What you will find of course is that you're frozen in place, for your raisin sits *stationary* with respect to the surrounding bread. And when you think about it a bit you realize that the very reason the raisins are moving away from you is because of the *expansion* of the bread. You and your raisin are not even moving; it's the space in between the raisins that is growing larger.

And that, precisely, is how we understand the cosmic expansion. Not as the movement of galaxies through an already existing, fixed, Newtonian space. No, it's much more interesting than that. The cause of the expansion of the universe is the space rushing into existence and flinging the clusters apart from each other. The size of each galaxy cluster stays the same, but the space *in between* the clusters expands in each instant, which results in an ever larger universe. That is our new understanding of the cosmos as a whole. A wild spirit breathes forth billowing chasms of space that explode the primeval fireball into a great growing immensity at whose center we find ourselves.

In terms of the large-scale expansion of the universe we are not moving. We are at the stable and unmoving center of this expansion. To be in the universe is to be at its center.

We are now in a position to show how the seeming paradox of being at the center and fifteen billion light-years from the center is in fact simply a counterintuitive feature of existence within an expanding universe. To see this imagine we are back near the beginning of time, in fact imagine we are right at the moment when the fireball begins to break apart and release its light in all directions.

Now, let's follow the adventure of a particle of light that is released very near to us. *If the universe were not expanding* such a particle would fly across the distance separating us in a matter of moments. But since the universe is not just expanding, but is expanding extremely rapidly, the particle of light has to travel a much greater distance as time passes. It's as if we were waiting at the top of the "down" escalator, and someone on the second step wants to reach us. The escalator begins moving very rapidly downward so that our friend, whose velocity never varies, is at first carried away from us. But as time passes the escalator begins to slow down so that eventually our friend makes it all the way back up the stairs and joins us at the top.

Just this happened with the photons of light released at the early moments of the universe. Those photons that were traveling in our direction were carried away from us by the rapid expansion of the universe. But as with a ball thrown up toward the sky whose initial speed upward slows down with every passing second, so too with the expansion of the universe, which began very rapidly but has been slowing down now for fifteen billion years. Those photons that were initially swept away from us kept traveling in our direction and eventually completed the journey to us.

The photons of light that arrived here in 1965 that were detected by Penzias and Wilson had been traveling toward us for fifteen billion years. So we can say that their place of origin *now* is fifteen billion light-years away from us. On the other hand, if we go back in time, we find that their place of origin *back then* is very close to where we are now.

Our own place here on Earth in the Virgo Supercluster was also an origin point of some of this primordial light, but we in the twentieth century do not see those particular photons. The matter we are composed of stayed *here* as the primordial light emanated away from us fifteen billion years ago. If there are intelligent beings elsewhere in the universe, they may be detecting those very photons of light that left from right here and that now arrive in their own distant planetary system with news of *our place* as the origin of the light from the beginning of time. We exist then at the very origin point of the universe, because every place in the universe is that place where the universe flared forth into existence.

12

Where Did the Universe Come From?

Now that we have arrived at the birthplace of the universe, we have one last question to explore, "If the universe began fifteen billion years ago in a blast of energy, where did all the energy come from?" We need to complete our journey to the center by approaching the power that gave birth at the center.

The discovery that we exist at the birthplace of the universe is monumental. It amounts to a cultural announcement that humanity is in the midst of a move to a new cosmological understanding of itself. How long this process of appropriating these truths and reinventing the human within this new and vast context will take is difficult to say because the truths are of a radical nature. To take our central example, an omnicentric universe seems to contradict what we have held as completely obvious about time and space and energy. But even before we can stop and assimilate the truth of omnicentricity we need to approach a discovery that is in some ways even more difficult to absorb than the omnicentric nature of our birth and evolution.

In all of this I think it helps to keep Einstein in mind. For if someone with such extraordinary conceptual powers had difficulty appropriating the new story of the universe, we can expect to have some of our own difficulties as we work to acquire understanding. The emergence of novel truths concerning cosmological questions is at times greeted with fear, joy, or sadness. Emotional intensities are inevitable, for there is no easy cerebral way to accept a revisioning of the universe in its most basic dimensions. I invoke the genius of Einstein once again as we confront one last twentieth-century discovery, that of the quantum vacuum.

As I have said earlier, one of the principal difficulties in understanding the new story is the Newtonian shape of our minds. This is nowhere more true than in the way modern understanding prejudices us concerning "the vacuum." For the modern mind, the "vacuum" means empty space. It means nothingness. It means "naught." And while there is a way in which such phrases can be considered true, we have discovered a deeper and more subtle dimension to the vacuum that we need to explore here.

Discussions concerning the vacuum sometimes point to the regions between the superclusters as the best approximation to a pure vacuum, and this is a reasonable way to proceed. Certainly matter and energy are extremely rare in between clusters of galaxies. But the unfortunate consequence of speaking in these terms is to give the idea that the vacuum is far away, and this is simply not true. The vacuum is everywhere, and the place I want to refer to in discussing the vacuum is the space right in front of you.

In order to bring the idea home, cup your hands together, and reflect on what you are holding there. What are the contents cupped by your hand? First in quantitative terms would be the molecules of air – the molecules of nitrogen, oxygen, carbon dioxide, and other trace gases. There would be many more than a billion trillion. If we imagine removing every one of these atoms we would be left holding extremely small particles such as neutrinos from the Sun. In addition, there would be radiation energy in the form of invisible light, such as the photons from the original flaring forth of the universe, or from Andromeda Galaxy and other sources. In order to get down to nothingness we would have to remove not only all the subatomic particles; we would also have to remove each and every one of these invisible particles of light.

But now imagine we have somehow done this, so that in your cupped hands there are no molecules left, and no particles, and no photons of light. All matter and radiation have been removed. No things would be left, no objects, no stuff, no items that could be counted or measured. What would remain would be what we modern peoples refer to as the "vacuum," or "emptiness," or "pure space."

Now for the news: careful investigation of this vacuum by quantum physicists reveals the strange appearance of elementary particles in this emptiness. Even where there are no atoms, and no elementary particles, and no protons, and no photons, suddenly elementary particles will emerge. The particles simply foam into existence.

I understand how bizarre and far-fetched this might sound for anyone learning it for the first time. But there is simply no way to make this discovery "reasonable." Most of us have Newtonian minds with a built-in prejudice that thinks of the vacuum as dead. If we insist that only material is real and that the vacuum is dead and inert, we will have to find some way to keep ourselves ignorant of this deep discovery by the physicists: particles emerge from the "vacuum." They do not sneak in from some hiding place when we are not looking. Nor are they bits of light energy that have transformed into protons. These elementary particles crop up out of the vacuum itself – that is the simple and awesome discovery. I am asking you to contemplate a universe where, somehow, being itself arises out of a field of "fecund emptiness."

The more carefully we study the universe, the stranger it gets. This emergence of particles out of a nonvisible field is not some unusual event taking place off in the regions between the superclusters of galaxies. This radical emergence takes place throughout the entire universe. The reason it took us so many millennia to discover this process is its subtlety. It takes place at a realm far more subtle than that which our eyes can detect. The usual process is for particles to erupt in pairs that will quickly interact and annihilate each other. Electrons and positrons, protons and anti-protons, all of these are flaring forth, and as quickly vanishing again. Such creative and destructive activity takes place everywhere and at all times throughout the universe.

The ground of the universe then is an empty fullness, a fecund nothingness. Even though this discovery may be difficult if not impossible to visualize, we can nevertheless speak a deeper truth regarding the ground state of the universe. First of all, it is not inert. The base of the universe is not a dead, bottom-of-the-barrel thing. The base of the universe seethes with creativity, so much so that physicists refer to the universe's ground state as "space-time foam."

Before we go further into this discussion, I want to stop a moment and reflect briefly on two of the reasons why this discovery will be especially difficult for us modern and post-modern people to comprehend. The first difficulty is linguistic, the second philosophical.

"Vacuum" as a word is truly pathetic as a vehicle for what we are intending here. At least in the English language, "vacuum" makes us think of a small, squarish device with hoses attached that are especially good for sucking up gum wrappers underneath our car seats. There's something terribly wrong about using a word with such homely connotations to refer to the basal generative power of the universe. I have suggestions for replacements, but before we discuss those let's consider the second, philosophical difficulty.

Significant portions of our modern, technocratic consciousness are based on the philosophical assumption that "reality" means "material things." We further assume, usually unconsciously, that the "really real" of the material object is its component part. The *whole* is seen as flimsy and derivative; the part as sturdy and primary. The Sun is taken as real, but the atoms that make up the Sun are even more real, for they are the building blocks that enable the Sun to exist. This reductionistic materialism also holds for our understanding of a living being. An animal is certainly real, but it is also vulnerable to death and decay in a way that its component atoms are not. Thus we suffer anxieties over whether or not our personal deaths will result in the extinction of our selves while remaining deeply certain that the reality of the atoms making up our bodies will continue long past our deaths.

Our reductionistic materialism is an outgrowth of a scientific tradition that held that the universe was built out of indestructible atoms, a tradition going back at least to Democritus in the classical Greek world. Certainly scientific understanding has benefited tremendously from this tradition. Because of it we were led to determine how every object in the universe is indeed composed of various atoms. This was far from obvious for earlier investigators and was in fact bitterly contested until it was empirically demonstrated. To discover that cumulus clouds, the Sun, and an ostrich are all composed of the same atomic elements is undoubtedly one of the greatest achievements of the human venture, and the atomic theory of matter is without question among the most robust in all of contemporary science.

The difficulty begins when we jump to the further philosophical assumption that reality is identical with these atoms. It was an easy mistake to make. If each thing is composed of atoms, why not consider the atom itself as the foundation of a thing's reality? This error was even easier at the beginning of the modern period when scientists such as Newton himself held that atoms were indestructible, eternal, the rock-bottom foundation of the universe. Eventually people came to think that to be a thing meant to be an aggregate of atoms, nothing more, nothing less.

To throw reductionistic materialism into doubt is going to be an upsetting experience for industrial society. I sometimes think the effects on contemporary culture will be similar to that experienced in an earlier period in human history when the idea of the divine rights of kings was thrown into question. People for centuries, even for millennia, had taken as obvious the right of the king's family to rule. The kings were thought to be either divine, as in classical Egypt, or directly descended from divinity, as in ancient India, or divinely ordained, as in medieval Europe.

Suddenly a new idea appeared suggesting that kings were

neither divine nor divinely ordained to rule, but that the real seat of political authority rested not with some genetically determined subset of humanity but with the entire people. Thus began the change in consciousness that led to the democratic revolutions transforming the world.

I realize that in comparing our reductionistic materialism with the divine rights of kings, I am comparing metaphysics with political philosophy and this may, at first glance, seem inappropriate. But the political implications of reductionistic materialism are real enough. To appreciate them we need only note the connection between our conviction that material is the foundation of reality and our devotion to consumerism.

If material stuff is understood to be the very foundation of being, we are quite naturally going to devote our lives and our education to the task of acquiring such stuff, for humans have an innate tropism for being. We move naturally toward that which we are convinced has greater being and value. Just as an earlier age devoted itself to serving the dictates of the kings in order to become directly involved with the really real, so too does our age dedicate itself to acquiring commodities in order to enter the wonderworld promised by our advertisers. To now suggest that material is not the only foundational reality in the universe throws some doubt upon one of the philosophical justifications of consumerism. And to begin doubting the unquestioned foundations of society's convictions is a dangerously creative and possibly destructive activity, as most monarchs, had they survived, would no doubt corroborate.

13

ALL-NOURISHING ABYSS

The true significance of the discovery of the quantum vacuum is the new understanding it provides concerning the reality of the *nonvisible.* I say nonvisible rather than invisible, for many things are "invisible" to us and yet are capable of being seen. Individual atoms are too small for the unassisted human eyesight to detect but such atoms can be seen if they are magnified sufficiently. The *nonvisible,* on the other hand, is that which can never be seen, because it is neither a material thing nor an energy constellation. In addition, the nonvisible world's nature differs so radically from the material world that it cannot even be *pictured.* It is both nonvisible and *nonvisualizable.* Even so, it is profoundly real and profoundly powerful. The appropriation of the new cosmology depends upon an understanding of the reality and power of the nonvisible and nonmaterial realm.

In contemporary physics the nonvisible realm is not pictured or given any sort of geometric form. It is rather depicted mathematically and is referred to with such words as "quantum fields," "quantum potential," "false vacuum," "possibility waves," "universal wave functions." For simplicity, I want to refer to this nonmaterial realm with a single phrase, and there are many possibilities to choose from.

We could simply use one of the phrases from physics, such as the "universal wave function." But the drawbacks with such a procedure are twofold. First is the question of misusing the languages of science. Science is not the same as cosmology, even when a cosmology is deeply informed by science. Cosmology is the story of the birth, development, and destiny of the universe, told with the aim of assisting humans in their task of identifying their roles within the great drama. Science on the other hand is focused on obtaining a detailed understanding of the physical processes of the universe. Language created for science was not created to work for cosmology, and to burden science's languages with this larger role would undoubtedly lead to undesirable ambiguities.

The second drawback to taking over science's languages for cosmological purposes is the unnecessary baggage of modern science's materialistic, mechanistic, and reductionistic bias. Even though this biased scientific exploration did eventuate in the discovery of this realm of nonmaterial power, the language of modern science will continue to be burdened by its history. To take over phrases directly from science is to risk confining the human imagination in the narrow corridors it haunted during the last couple of centuries while today the very discoveries of science cry out for a much richer and deeper awareness.

A second way of identifying the ground of the universe would be to take not science's language but theology's, but this would be just as problematic. Certainly it is true that humans within classical traditions have reflected upon generative possibility for millennia now, and though this meditation was based on modes of inquiry different from modern science's, it brought forth insights of profound significance in the quest for understanding. Instead of speaking of the particles as foaming forth from the universal wave function or the quantum vacuum, such theologians would speak of the generative powers of the Logos, or the Tao, or Shunyata.

Cosmology is not the same thing as theology; as before, in the case of using science's words, there would be the question of misusing theological languages. Theology is the rational inquiry into the nature of God and humanity's relationship with God. Theological terms of discourse were not invented to tell the story of the universe, and to force theology's terminology into the service of cosmology would almost certainly create a lot of unnecessary confusions.

A second and equally serious difficulty with gathering up theology's terms for telling the new story of the universe is the residue laced into modern theology. During the modern period much of theology focused on the unique aspects of the human-divine relationship, and this tended to give theology a subtle and sometimes outright bias against nonhuman nature. The great news of our time is the evolutionary story in which we come to realize that we humans are all embedded in a living, developing universe, and that we are therefore cousins to everything in the universe. To employ theological language emphasizing our separation from the universe is to burden our endeavor with unnecessary baggage.

Cosmology as an ancient wisdom tradition draws from science, theology, art, poetry, and philosophy, but is, strictly speaking, its own distinct tradition. It's not a question of eschewing scientific or theological terms altogether, but rather of aiming for language arising out of our experience of living within an unfolding cosmos.

I use "all-nourishing abyss" as a way of pointing to this mystery at the base of being. One advantage of this designation is its dual emphasis: the universe's generative potentiality is indicated with the phrase "all-nourishing," but the universe's power of infinite absorption is indicated with "abyss."

The universe emerges out of all-nourishing abyss not only fifteen billion years ago but in every moment. Each instant protons and antiprotons are flashing out of, and are as suddenly absorbed back into, all-nourishing abyss. All-nourishing abyss then is not a thing, nor a collection of things, nor even, strictly speaking, a physical place, but rather a power that gives birth and that absorbs existence at a thing's annihilation.

The foundational reality of the universe is this unseen ocean of potentiality. If all the individual things of the universe were to evaporate, one would be left with an infinity of pure generative power.

Each particular thing is directly, and essentially, grounded in all-nourishing abyss. Though we think of our bodies as dense and completely filling up the space they occupy, careful investigation of matter has shown that this is not the case. The volume of elementary particles is extremely small when compared to the volume of the atoms that they form. Thus, the essential nature of any atom is less material than it is "empty space." Even from this elementary perspective we can begin to appreciate that the root foundation of any thing or any being is not the matter out of which it is composed so much as the matter together with the power that gives rise to the matter.

All-nourishing abyss is acting ceaselessly throughout the universe. It is not possible to find any place in the universe that is outside this activity. Even in the darkest region beyond the Great Wall of galaxies, even in the void between the superclusters, even in the gaps between the synapses of the neurons in the brain, there occurs an incessant foaming, a flashing flame, a shining-forth-from and a dissolving-back-into.

The importance of the cosmological tradition is its power to awaken those deep convictions necessary for wisdom. Knowledge of all-nourishing abyss is the beginning of a process that reaches its fulfillment in direct taste. We think long and hard about such matters as a way of preparing ourselves for tasting and feeling the depths of a reality that was always present and yet so subtle it escaped us.

It may be that in the next millennium religious convictions will be awakened and established within the young primarily by such meaningful encounters with the mysteries of the universe, and only secondarily by the study of sacred scriptures. The task of education then will focus on learning how to "read" the universe so that one might enter and inhabit the universe as a communion event.

As one brief illustration of what I mean here, I'd like to consider the Moon at night and to indicate how, within the new cosmology, the Moon can become an activator of those deep convictions necessary for wisdom.

Before we had an understanding of the quantum nature of the universe we could so easily think of the Moon as just this object in the sky. Its light was thought of as just the light from the Sun that had been reflected our way. And of course there's some truth here, for if the Sun were suddenly to go dark then the Moon as well would go dark. But there is another and more subtle quantum sense in which the Moon is not just an object and its light is not just reflected from the Sun.

Newton and others early in the modern period regarded the Moon as a conglomeration of stable, unchanging atoms. With such a conception it is easy to hold that the Moon simply forms a wall off of which the light from the Sun bounces. But when we examine the physical processes of "light," "atoms," and "bounce," we find a much more complex and even astounding dynamic.

First, the elementary particles and atoms are not permanently existing objects but are events that are vibrating at extremely rapid rates. Even the word "vibrate" is not exact, for it connotes a solid object that moves rapidly back and forth in space. This gives a false image of what our data indicate, for we know in fact that it is not true to think of the particles as moving back and forth in space. Rather, as has been celebrated and discussed throughout most of the twentieth century, particles exist in one location and then exist in another location *without traversing the space in between*. So, as bewildering as it might sound to us, it is more accurate scientifically to say that the particles and atoms are flashing into existence, surging into existence, and then just as suddenly they are dissolving from their place to surge forth in a nearby location, all happening so rapidly that the unassisted human eye cannot catch the movement.

The Moon is not a dead object, but is instead an ongoing scintillating event.

Second, it is false to think of photons as "bouncing" the way a ball would bounce when thrown against a wall. Instead, the photons from the Sun "interact" with the particles of the Moon. As with every interaction at the quantum level of reality the process of this interaction begins with the annihilation of the particles as they are absorbed into the all-nourishing abyss and is followed by the creation of a new set of particles. If this new set contains any photons, these photons are *new.* They did not exist in the previous instant but, rather, came forth out of the annihilating event of the interaction.

Thus, it is not true to say that the photons of light arriving here from the Moon have just been bounced from the Sun. Moonlight comes from the Moon, for moonlight is *created* by the Moon.

In the next millennium, young people educated in the new cosmology will experience the Moon not as a frozen lump but as an event that trembles into existence each moment. Moonlight will be understood not as bounced from the Sun but as *expressive* of the Moon's reality. They will regard the Moon not as dead object but as a creative source, as an origin of the universe, a geyser in the sky where the universe sprays into existence.

Through such encounters we learn that the universe is not a collection of dead objects but is, rather, a seamless whole community made up of cosmos-creating subjects.

When children learn of the universe's birth, they ask, "What was before?" These minds of ours, emerging fifteen billion years after the great flaring forth, these minds of ours – woven tapestries of the same primal particles emerging in the beginning – these minds of ours insist upon knowing what is at their own base. We wish to know the nature of the reality from which we arose, for then we will know our own deepest nature.

Humans of every culture have contemplated this mystery, and we in the twentieth century now enter the cosmological lineage with our own contribution. We too have contacted, in our own unique way, the Great Power that gave birth to the universe fifteen billion years ago and that continues to be involved with giving birth in every interaction throughout the universe today. From our own fresh empirical-mathematical-observational contemplation of the universe we have identified a nonmaterial realm suffusing not only the great macrocosm of the universe, but suffusing just as thoroughly the microcosm of the human and of every being of the Earth and universe. That which gave birth to the universe is giving birth in this moment as well. Although our understanding is very young, and thus inadequate in many ways, what we have discovered is already profoundly stirring.

From our cosmological studies we learn that we are at the center of the large-scale omnicentric expansion of the universe. From our quantum studies we learn that matter and energy emerge from the quantum potential. Taken together these two discoveries bring human understanding to a vista never before enjoyed: Each child is situated in that very place and is rooted in that very power that brought forth all the matter and energy of the universe.

14

EINSTEIN'S AWAKENING

After all the theoretical discussion on the origin of matter, as well as my linguistic anxieties about what words to use in order to name the origin, I want now to turn to something concrete, to a person, to a specific moment in time in order to ground this whole discussion. My hope is that if we can understand these ideas as they pertain to an actual human, we will have a better hold on how they shed light upon our own historical personal situations.

So we turn one last time to Albert Einstein. It is not Einstein the person that I am primarily interested in; it is Einstein as an image or a revelation of what it means to say that our lives are simultaneously our own ordinary lives and the central creativity of the evolving universe. The challenge is in identifying our deepest personal reality with the powers of the universe, and in order to reflect further on this I want to return to that archetypal night when Einstein was hunched over his mathematics in his second-story Berlin apartment.

So there he sits, this human, his frame bestowed by the primates and now held aloft twenty inches by rented furniture. Even if no actual drums are playing as Einstein contemplates the structure of the universe, certainly from our later understanding of the significance of this moment we can imagine that in the dark night there thundered a kind of silent music in celebration of what was here taking place.

A strange moment certainly and yet so much that was familiar too. His pipe leaning on its side, forgotten, the ashes dead for an hour now. The room heavy with the sweet, sharp smell of the old smoke. Outside, raining still, the few people on the streets below hurrying under caps and dark coats. But Einstein hasn't looked out the window all evening, and even if he did the sensory images would hardly penetrate his consciousness. Not aware of the rain, or the dead pipe, or the color of his dark green pencil, yet profoundly aware, and very alert, even heavy laden with consciousness, a sort of consciousness that lives best near the edge of the world, a region he seemed to have occupied so often.

Psychological investigators later questioned Einstein about his consciousness during such moments of radical creativity, and he stunned them with his reply. They had been curious about Einstein's mathematical orientation. Most mathematical scientists fall into one of two qualitatively distinct mental domains. One is the algebraic, in which an investigator will rely on the formalisms of equations with alphabetic letters and algorithmic rules for transforming one symbol set into another. The other is the geometric, where, instead of using abstract formulas, an investigator relies upon shapes and spatially constructed scenarios to represent the dynamics. After interviewing a number of world-class mathematical scientists, the investigators had grown accustomed to categorizing their subjects under one of these two domains.

Instead of speaking of algebraic formalisms or geometric

intuitions Einstein, after reflecting on the question carefully, said that his experiences were of "the muscular type."

As Einstein sat, thick with awareness, his hands pressing the mathematical symbols onto the page, the Hercules Cluster of galaxies was racing away from the Milky Way at thirty thousand kilometers each second. The Coma Cluster of galaxies too, all its trillions of stars each a million times the size of Earth, sailed silently into the cosmic night. There was no awareness of this in the Earth's mental activities, and Einstein sat at brood.

Fifteen billion years earlier the universe flared forth as stupendous energy, and in this primordial fire of the beginning each point was the unwobbling center of the expansion. Through the numinous alchemy of the creativity pervading the cosmos some of the first particles there at the center baked into a chunk of Earth that sat meditating, protected from the rains of self-contempt, or financial insecurity, or needless distraction.

Several times each year the Milky Way gives birth to a new star. As Einstein sat there and the distant galaxies raced away, the Milky Way was busy in all regions of its domain, nurturing clouds of elements that sat brooding with the potentiality of star birth. I wish I could write with certainty that simultaneous with Einstein's awakening to the dynamics of the large-scale structure of space-time a star was born in one of the arms of the Milky Way. But even if this simultaneous birth did not take place, it is nevertheless true that in both such births we have a patch of the Milky Way, thick with the creativity of the Milky Way, churning with urgency, approaching new birth out of the multivalent potentiality of the Milky Way.

Einstein eventually despaired of ever explaining his experience of creativity to others. So many suffered under a distorted understanding. He did not, he told them patiently, simply study data and then look for equations that fit the data. The truth of the matter, so difficult for them to accept, was that he himself relied primarily on imagination.

When Einstein – this piece of the Milky Way – was asked what he sought, he answered: "I want to know how the Old One thinks. The rest is a detail." This creature wearing worn-out shoes, this mustachioed member of the notochord phylum, this living flesh with its creases from a lifetime of laughter, this lumpy concentration of molecules, this soul searing with wonder burned to know how the Old One thinks. How the Old One is shaping the great vaults of the heavens. How the Old One is casting a billion stars in their circlings. How the Old One is fastening the baryons together, how the Old One is releasing electromagnetism throughout the cellular membranes.

Chock-full with the very dynamics he sat contemplating, Einstein experienced a birth that permeated him whole, his mind, his muscles, his viscera. Effortlessly, and as a form of these very dynamics, he jotted down the field equations. This chunk of the Milky Way jotted down the dynamics of the Milky Way. This region of space-time rich with the interactions of the universe jotted down the symbolic form of the interactions of the universe. This fleshy portion of the world transformed its insides into graphite to reveal the harmonies at work throughout the fleshy world.

The Indians of South America teach that to become human "one must make room in oneself for the immensities of the universe." Unless we do so, we cannot find our true nature. We will wander in pain and loneliness. We will never learn how the Old One thinks. Caught in fragments of our nature, we will attach ourselves to one fragment after another, each taking us further away from our center.

Making room for the immensities, Einstein experienced their inrush, when suddenly the Milky Way as Great Self became this Einstein reflecting upon his deepest nature, which is the nature of the galaxy and the cosmos too. Afterward, when he had calmed down and was attempting to make sense of what had happened, he did distort some of what had surfaced. He added a symbol. He changed what he had intuited. In the cultural calm away from the thundering breakthrough he gave in to the habits of his time and place and altered his original equations – a shrinking back that we can reflect on to our benefit.

But in that great moment, in that state of consciousness coming from years of disciplined preparation, *Einstein was not contemplating something apart from himself.* He was absorbed in the experience of the feelings in his body, his viscera, his bowels, that were caused by the causes permeating the universe. That Great Power that had, there, at the birthplace of the universe, gushed forth in all the energies and galaxies was now bringing forth its own self-portrait in the symbols of Einstein's field equations. Powers that would one day receive such names as "gravity," or "the second law of thermodynamics," or "the strong nuclear interaction"; powers that shaped the Milky Way Galaxy and the mammalian organism; those were at work in that concentrated form of the Milky Way called Einstein, and it was there that this Great Power broke into a new contemplation of its sublime grandeur.

15

THE CENTER OF THE COSMOS

The universe began as an eruption of space, time, matter, and energy out of all-nourishing abyss, the hidden source of all creativity. The universe began as a titanic bestowal, a stupendous quantum of free energy given forth from the bottomless vaults of generosity.

The nature of this original gift goes so far beyond our daily human experiences that we must resort to numbers to approach it. In the first second, the universe is a million times hotter than the center of our Sun today. It is in an extremely compact form. The density of the original matter is a billion times greater than rock. And yet this primordial matter is exploding forth as rapidly as the speed of light.

Every place in the universe is at the center of this exploding reality. From our place on Earth today in the midst of the Virgo Supercluster, all of the universe explodes away from us, just as it does from the perspective of anyone in the Perseus Supercluster. We are at the unmoving center of this cosmic expansion, and we have been here at the center from the beginning of time.

The universe began here in a different form, one so hot no structures could yet exist, but as the expansion continued, the temperature slowly came down, and the first assembled beings began to appear. Quarks, the constituents of the stable elementary particles, gathered together and formed protons and neutrons. Three minutes later these in turn formed the first nuclei. After three hundred thousand years the temperature dropped to six thousand degrees, that of the surface of our Sun today, and the universe transformed itself from nuclear particles to the first atoms of hydrogen and helium. This same spectacular transformation continued into the future, carrying these atoms into the form of the galaxies, and then into that of the molecules and cells, and then into the very form of the human and the elephant and the blue spruce and the Mississippi River.

That which blossomed forth as cosmic egg fifteen billion years ago now blossoms forth as oneself, as one's family, as one's community of living beings, as our blue planet, as our ocean of galaxy clusters. The same fecund source – then and now; the same numinous energy – then and now.

To enter the omnicentric unfolding universe is to taste the joy of radical relational mutuality. For we know this body of ours could have been a giant sequoia. We know in a simple and direct way that we share the essence of and so easily could have been a migrating pelican. Our astonishment at existence becomes indestructible, and we are home again in the cosmos as we reach the conviction that we could have been an asteroid, or molten lava, or a man, or a woman, or taller or shorter, or angrier, or calmer, or more certain, or more hesitant, or more right or wrong.

I have given some scientific detail in these chapters as a way of grounding this new cosmology in science's empirical and theoretical discoveries, but here at the end it is important to understand that the center of the cosmos is not "mathematical science," nor is it "inside" science, nor is it "owned" by science. *The center of the cosmos is each event in the cosmos.* Each person *lives* in the center of the cosmos. Science is one of the careful and detailed methods by which the human mind came to grasp the fact of the universe's beginning, but the actual origin and birthplace is not a scientific idea; the actual origin of the universe is where you live your life.

The primary challenge of this cosmological transformation of consciousness is the awareness that each being in the universe is an origin of the universe. "The center of the cosmos" refers to that place where the great birth of the universe happened at the beginning of time, but it also refers to the upwelling of the universe as river, as star, as raven, as you, the universe surging into existence anew.

The consciousness that learns it is at the origin point of the universe is itself an origin of the universe. The awareness that bubbles up each moment that we identify as ourselves is rooted in the originating activity of the universe. We are all of us arising together at the center of the cosmos.

INDEX

advertising, 13–19, 32, 57
alcohol, 33
alienation, 33–36
all-nourishing abyss, 100–101, 103, 110
Andromeda Galaxy, 60–62, 64–65
Aristotle, 39
atomic theory of matter, 95
atoms, 94–95, 100, 111
awe, 47–48

birthplace of the universe
  communicating, 2–7
  discovery of, 2, 64, 66
  distance from, 63, 78, 83–84
  radiation emanating from, 78–79, 83, 88–89, 107. *See also* photons

Carson, Rachel, 18
cave paintings, 8
chimpanzees, 49–50
clusters, 75
Confucius, 18, 39
consciousness
  altering basic patterns of, 49
  classical, 78
  mythical, 78
  scientific, 78
  transformed, 45
consumerism, 15–19, 32–37, 57, 96
Copernicus, Nicolaus, 21–23, 38–39, 80, 83
cosmic egg, 76, 79, 111
cosmological constant, 72
cosmology
  advertising as, 13–19
  aesthetic and affective dimensions of, 32
  defined, 7, 8, 98
  and drugs, 33–36
  task of, 9, 31
  as wisdom tradition, 31, 99
cults, 19
culture, 9–10
Curie, Marie, 72

Darwin, Charles, 72
Democritus, 94
divine rights of kings, 95–96
Draco, 61
drugs, 33–36

Earth
  as center of the universe, 21–22, 80, 83–84, 104
  as revolving around the Sun, 27–28
  size of, 28–29, 39, 53
Earth Community, 3, 4, 34, 56
ecstasy, 32, 36–37
education
  advertising as, 13–19
  cosmological, 32, 39, 86
  Ecozoic, 58
  modern system of, 32
  moral, 44
  task of, 101
Einstein, Albert, 6, 38, 64, 70–75, 81, 86, 90, 91, 105–9
English language, 29

faith, defined, 17
false vacuum, 97
Feynman, Richard Phillips, 38
field equations, 74, 75, 108, 109
Fornax, 61
Friedmann, Alexander, 73

galaxies, 45–46, 61–63, 75, 81, 84
Galileo, 39, 72, 83
Gamow, George, 75, 77, 78
General Theory of Relativity, 71
genetic codes, 9, 50
gravity, 28, 52
Greeks, ancient, 39–40, 94

Hercules Cluster, 86
Herschel, William, 72
Homo sapiens, 5, 6
Hubble, Edwin, 63, 64, 68, 73, 75, 80, 83, 84, 86

initiation, cosmological, 13–19, 36. *See also* education
isolation, 33–36

Jesus, 18
joy, 36, 111
Jung, Carl Gustav, 40, 73
Jupiter, 26, 27

Kant, Immanuel, 67
Kepler, Johannes, 72
King, Martin Luther, Jr., 18

light, 39. *See also* photons, Sun
light-year, defined, 59
Local Group, 62
local universe test, 56
loneliness, 33–36

M32, 61
McClintock, Barbara, 38
Magellanic Clouds, 61
magicians, 36
Mars, 26
materialism, reductionistic, 94–96
mathematics, 65
mental domains
  algebraic, 106
  geometric, 106
  muscular, 107
Michelangelo, 21
Milky Way Galaxy, 47–49, 53, 58–62, 75, 107
Moon, 101–3
"muscular" experiences, 107
myths, 39–40

Neolithic cultures, 76
new awareness, 29
new story, 25
Newton, Isaac, 38, 72, 95, 102
Newtonian world-view, 70–71, 84–87, 91, 92
Nietzsche, Friedrich, 83
night, 67–68
Nintendo, 57
nonvisible, 97

Oedipus Rex, 55, 56, 60
Old One, 108–9
omnicentricity, 85–87, 90, 104
orangutan, 50

Palomar, Mount, 73
Penzias, Arno, 78, 83, 89
perception, changing, 24–31, 45
photons, 1, 59, 64, 65, 78, 79, 88, 89, 103
photosynthesis, 40
Plato, 39
poetry, 39, 42
possibility waves, 97
primal peoples, 53
primates, origin of, 25

quantum fields, 97
quantum potential, 97
quantum vacuum, 91–94, 97
quarks, 111

reductionistic materialism, 94–96
Relativity, Theory of, 71

religion
  celebrations in, 10
  consumerism as, 13–20
  and science, 11–12
  value of, 12
Robertson, Howard, 68
Russell, Bertrand, 83

sacred stories, 9
sacrifice, 40–44
Sagittarius, 58
Sartre, Jean-Paul, 83
science
  aims of, 31, 98
  beginnings of modern, 21
  as collective enterprise, 73
  as counterintuitive, 24
  domain of, 11–12
  hypothesis in, 77
  irruption of, 3
  languages of, 98
  method of, 77
  nature of, 76–77
  purpose of, 1
  and religion, 11–12
  as wisdom, 3
scriptures, classical, 11
Sculptor, 61
Sega, 57
self-awareness, discovery of, 4–5
shamans, 36
Sitter, William de, 68
Skydome (Toronto), 46, 48
Slipher, Vesto, 63, 68
Socrates, 18
Sophocles, 55
Soviet Union, 13, 15
space, birth of, 85
space-time foam, 93
Special Theory of Relativity, 71
star birth, 107
Sun, 27
  as center of the solar system, 21–23, 38–44
  gravitational power of, 29
  as self-giving, 42–44
  size of, 28–29, 53
  transformed into energy, 39–40
sunset, 23–24, 26–27
superstitions, 10
telescopes, 65
television, 15, 18
Thebes, 55, 56
theology, 98–99
time, birth of, 85
tools, 2, 23, 61
truth, 6

unhappiness, 17
universal wave function, 97, 98
universe
  expanding, 71–76, 87–88, 110
  explosion at origin of, 77–78, 84
  ground state of, 93
  See also birthplace of the universe

vacuum
  false, 97
  quantum, 91–94, 97
Venus, 26–28
videogames, 57
Virgo Cluster, 62, 89

water, 10–11
Wilson, Robert, 78, 83, 89
Wirtz, Carl, 68
wonder, 47–48